D0901236

Cable TV Renewals & Refranchising

Edited by Jean Rice

Cable TV Renewals & Refranchising

Edited by Jean Rice

Executive Editor
Mary Louise Hollowell

Communications Press, Inc.
Washington, D.C.

Grateful acknowledgment is made for permission to print the following: the Appendix "Requesting Proposals for Cable Television Systems: A Model for Municipal Governments," © 1982 by the National Federation of Local Cable Programmers (NFLCP); and the Glossary, compiled by Mary Louise Hollowell, © 1980, 1983 by Communications Press, Inc.

ISBN 0-89461-037-6

Published by Communications Press, Inc.

Printed in the United States of America

Cover Design by Linda McKnight

Contents

Jean Rice

1. Renewals and Refranchising: Municipal Options and Procedures

THE EVOLUTION OF CABLE

Cable television has grown dramatically in the last twelve years—from 7.6 percent of the TV homes receiving cable in 1970 to the 34 percent being served in 1982. Another major increase in the number of wired households is expected over the next five years as cable continues to move to the major urban areas. The five year flurry of franchising in metropolitan and surrounding areas was coined "The Last Gold Rush." With only a few major markets left to be franchised, attention is now focusing on franchise renewals and refranchising. The expiring franchises were granted to cable companies by municipal officials in the 1950s, 1960s, and early 1970s. Many of the early franchises ran twenty-five years, with more recent contracts terminating in fifteen years.

Thirty years ago cable television was a transmission medium which provided improved television reception. Initially, a cable was strung from an antenna placed on the highest point of local terrain to homes in a valley. The desire for improved reception of broadcast television signals was what

---

Jean Rice is President of Rice Associates, Inc., a consulting firm to cities, specializing in franchising, refranchising/renewal, and alternative ownership. Previously, Ms. Rice held positions as Telecommunications Management Specialist, National Telecommunications and Information Administration (NTIA), U.S. Dept. of Commerce, and Senior Municipal Consultant, New York State Commission on Cable Television.

spurred the early development of cable.

Cable television as a transmission medium and a business has changed in technical capacity, consumer offerings, operations, and financial viability since the early days. In technical capacity 36- and 54-channel cables have become the norm, with some urban areas being served by dual cables (offering up to 104 channels) with additional cables serving institutions and businesses within the community. New cable systems on an almost routine basis provide "upstream" channels which offer the technological capacity for a variety of two-way services such as traffic signal control and fire alarm surveillance. Systems also offer addressability—the ability to transmit programming to specific homes.

Production equipment for local programming has also advanced rapidly from the offerring in the early 1970s of low cost portable black-and-white equipment to color equipment now being offered for the same price. (See Chapter 2 for a more detailed description of technological changes.)

Not only have changes in cable systems and production equipment advanced the state of the art, but satellites have produced a new low cost program delivery mechanism. Cable systems, prior to satellite delivery in the 1970s, received programming primarily from aerial antennas and terrestrial microwave networks. While programming is still received by these traditional methods, satellites provide distant broadcast signals, pay TV, and more to large geographic areas at low cost. Cable systems now build between one and three satellite receivers to capture satellite delivered programming.

The expanded technological capacity of cable systems, the advent of satellite delivered programming, and low cost production equipment have stimulated new consumer offerings. Today not only does cable offer the retransmission of broadcast signals, but specially designed program packages (e.g., childrens, movies). The systems can offer subscribers some of the over 50 program services available. System operators package the entertainment services that they feel will be of interest to local residents. (Chapter 2 also contains a broader description of the programming packages systems offer.) Local cable programming is now provided in color and in new systems, three to ten or more channels are set aside for this purpose. (See Chapter 6 for an in-depth discussion of local programming.)

Non-entertainment interactive services are also on the increase. In larger systems one or more of the following services are being offered: push button polling, traffic signalization, data transmission, energy management, and security and fire alarm services. On an experimental basis videotext services and transactional services—shopping and banking at home—are being offered.

Entertainment and non-entertainment consumer services have grown tremendously since the early 1970s. There is, however, still a great deal of flux in this part of the cable industry. Some new services have failed, such as the CBS arts channel; some have succeeded, such as Home Box Office; while others are still in the developmental stage. The economic viability of many of the new interactive services is just now being tested. Energy management systems have been saving an average of 15-20 percent in fuel consumption, and fire alarm systems have proven effective. The speed and actual development of text and other services will depend on the outcome of existing experiments and the willingness of a variety of companies to develop new products and services.

In system operations, cable operators have gained sophistication in customer service and marketing. Customer service has been enhanced by internal computer systems for consumer billing and complaint handling. The utilization of addressable systems for changing tiering compositions requested by subscribers has also enhanced operational efficiency. In the last four years marketing has been one of the major topics of industry seminars with the tiering of services (e.g., basic—including local and distant TV signals and some satellite services, pay-movie, and other specialized programming packages) a major topic. New marketing techniques, such as printed program guides and pay service informational brochures, have been developed. These operational changes have helped in increasing initial and sustained subscribership.

Cable system capacity increases, marketing improvements, cost efficiencies derived from utilizing satellites to transmit programming, and consumer acceptance of pay programming packages have been major factors in the increased financial viability of the cable industry. Industry revenues have virtually doubled since 1975 with satellite delivered pay services being the predominant source of new revenue. The increased revenues have led major companies to invest in cable and financing institutions to provide loans for cable development.

As a whole, the industry has evolved substantially from the mom and pop operations that strung wires from mountain tops to provide themselves and their neighbors with better TV reception. Today, the typical owner is a multiple systems operator (MSO). Systems have changed from offering few channels to many with vastly increased consumer offerings. Financing for cable, even in this recessionary period in the 1980s, is strong. Municipalities whose franchises are up for renewal or are expiring need to become informed about the industry as a whole with an emphasis on how these changes could impact the service offered local residents.

THE BASICS OF RENEWAL AND REFRANCHISING

The basic document in renewal and refranchising processes is the franchise. The franchise agreement provides the cable system with authorization to utilize public rights of way in conducting business and establishes the terms and conditions under which this can be done.

Although franchises in some locations are awarded at the county or state level, the vast majority are awarded by local jurisdictions. The method of award varies between jurisdictions but typically is done in one of the following manners: through a municipal ordinance which includes the terms of the franchise; through a municipal ordinance which references a separate franchise agreement; and through other existing municipal procedures (e.g, licensing). Since the purpose and legal status of ordinances and franchises are generally the same, the terms are used interchangeably in this book.

Renewal of the franchise and refranchising can be accomplished in a number of ways. The first and most often adopted procedure is renewal either through an "automatic" renewal or renegotiation of the incumbent cable operator's franchise. An automatic renewal of the franchise usually occurs when the expiring franchise requires automatic renewal or renewal at the option of the operator. Renegotiation with the existing operator is undertaken for two major reasons. First, the offering of additional channel capacity and new services can be made in a phased approach that most easily avoids the interruption of service to local residents. Second, if a system upgrade would meet community needs, this can be done more cheaply by the existing operator than by a new operator.

One of the major disadvantages of renegotiation is that renewed franchises may provide less channel capacity, consumer services, and access support than a competitive refranchising procedure would elicit. Additionally, existing operators have been unwilling to pass on the savings incurred by upgrading rather than rebuilding cable systems and have not offered cities the same packages routinely provided in initial franchising today.

Refranchising takes two forms, competitive refranchising and alternative ownership. In the competitive refranchising process, bids are solicited from a number of companies, usually including the incumbent. Municipalities usually pursue this route if they are dissatisfied with the company performance or feel that a sufficiently competitive market exists to elicit a better service package than could be obtained through renegotiations. This approach while being undertaken by an increasing number of cities is not the norm primarily because cities felt that renewal provided for a quick and easy way to proceed and because the competitive environment was not sufficient to warrant a refranchising approach. The major advantage to refranchising is that competition spurs a higher level of service offering than has been garnered through renewal processes.

Refranchising through alternative ownership is also not common practice but is being increasingly considered by municipalities. The principal reasons people consider municipal or cooperative ownership are dissatisfaction with the existing operator performance and the belief that lower cost and better quality service can be provided. Municipalities taking this approach to refranchising feel that the service package benefits of alternative ownership outweigh those to be garnered either through the competitive process or renewal.

Each municipality must weigh the pros and cons of the renewal and refranchising approach and select the one appropriate for their situation. In some cases, renewal proceedings may be started and if unsuccessful, refranchising procedures can be initiated. The methods utilized will directly impact the quantity and quality of consumer and municipal service. The key for municipal officials is to be informed on the state of the cable industry, the market of the existing system, and what reasonable expectations for service would be.

MUNICIPAL CONSIDERATIONS

As discussed, the municipality has several procedural options to pursue as franchise expiration and renewal approaches. There are many considerations to be taken into account prior to deciding on which procedure to pursue. Following is a brief description of some of the relevant issues and goals.

Company Performance

One of the first elements municipalities consider is performance of the incumbent company. This includes responsiveness to consumer complaints, ordinance compliance, prompt response to municipal requests, technical quality, area and timeliness of construction, financial stability, marketing ability, local programming contributions, and other aspects of daily operations. From this review and past relationships with the company, municipalities can determine if they have been well served. Some cable systems, in communities where reception is poor, subscribership high, and 12-20 channels are offered, are considered the "cash cows" of the industry. In these situations, financial reporting will provide information to the municipality which will help determine if the system has been upgraded when feasible, if rates are justified, what the rate of return has been, and other elements. (A discussion of the cable operator's perspective on renewal can be found in Chapter 5.) This information will assist the municipality in determining which procedures to follow. Municipalities that feel they have received quality service are more likely to pursue the renegotiation or automatic renewal approaches.

Competitive Environment.

Prior to a municipality opting for refranchising, an analysis of the competitive environment is useful. In the case of very small communities, it may be that no other operators would be willing to serve the area. In other cases, large multiple system operators (MSOs) have maintained a code of ethics which predisposes them not to compete with one another during the franchising process. This code of ethics though prevalent today may change in the future as the room for expansion in new markets decreases and MSOs have completed construction in major urban areas. Small cable systems and small MSOs have shown a decided interest in competitive refranchising. If the market is a lucrative one, cities which have tested the environment have found it competitive, though not as competitive as would have been expected in the height of the franchising boom.

Ownership Patterns

Ownership patterns in cable have changed dramatically from an industry of "mom and pop" operators to an industry of MSOs. The top twenty cable companies hold approximately 51 percent of all cable subscribers in the country. Not only is there a trend towards concentration of ownership among companies with existing media interest. In 1970, 49 percent of the cable systems were owned by media related companies; in 1980, this increased to 72 percent. So not only have concentration trends appeared—but cross ownership with other media and vertical ownership have appeared as well. Vertical integration trends arise as media companies own program production facilities, satellite distribution capability, and the final mile-cable. In some instances this has led to companies precluding the entry of other pay services on their systems. These ownership trends have raised questions on the ability of subscribers to have a variety of programming from diverse sources, on the efficiency of single concerns dominating the media and information to the home, and on the ability for such large systems to be responsive to local needs and concerns. A trend of increased minority ownership which was heralded in the early 1970s never materialized and municipalities are concerned about the no to low minority ownership pattern that has developed. Municipalities have responded to these various concerns by setting their own limitations on crossownership and by promoting local and minority ownership, and the availability of leased access channels.

Deregulation

For four consecutive years, bills have been offered in the U.S. House and Senate which would deregulate cable. The bill up for consideration in 1982, S.B. 2172, would substantially deregulate the cable industry, thereby providing little room for local negotiations and regulation. This bill would require automatic renewal for cable systems in substantial compliance with franchise agreements. Since the fifteen- to twenty-year old documents may easily be "substantially" complied with, the municipality would not have the ability to negotiate for improved system capacity and service. Additionally, a small maximum of channels for public sector and leased access would be set and new refranchise contracts could not request or require additional channels for this purpose. Rate regulation would be relegated to a limited portion of the cable system, off-air broadcast signals and access. Regarding alternative ownership, it could be accomplished in the refranchise-renewal context if the systems were purchased for fair market value. Purchase at fair market value typically means that the municipality would be paying for the use of its own right of way. Municipalities and national organizations such as the National League of Cities and the National Federation of Local Cable Programmers have opposed this legislation while MSOs and the National Cable Television Association have supported it. This strong deregulatory effort has left many municipalities wary of their future ability to regulate and has catalyzed consistent federal lobbying efforts. (For further discussion of this bill see *The Cable/Broadband Communications Book, Volume 3, 1982-1983.*)

Alternative Ownership

An increasing number of cities in the franchising, refranchising, and renegotiation processes are considering alternative ownership. Reasons given include: lower consumer rate structures; increased revenues for the city; ability to utilize revenues for social and economic development; ability to ensure public sector and leased access availability; private company past performance; federal deregulatory attempts; local ownership and control;

and provision of job opportunities. The most common approaches are municipal ownership and cooperative ownership. Most municipalities considering this option are viewing cable in terms of its long term benefits, not short term financial gain. Through the establishment of municipal commissions and non-profit access corporations, cities pursuing municipal ownership can actually advance rather than detract from First Amendment objectives. Cooperatives have the ability to meet subscriber objectives without requiring the same return on investment that private companies require. While industry representatives note legal complications in pursuing alternative ownership options, municipalities undertaking new systems, like the cooperative in Davis, California, and the municipal system in Wyendotte, Michigan, have not found legal impediments.

Rural and Low Income Service

Providing service to rural and low income areas is a major consideration for many municipal officials. In some cases low income and rural areas were not built during the initial franchise term. The issue in rural development has always been the cost factor. A new technique pioneered by the Rural Electrification Administration shows ways to cut these costs. Chapter 7 addresses this in more detail. Chapter 8, which discusses low power television, also addresses this problem.

In providing service to low income residents, construction in the area is a major consideration and can be likened to normal construction timetables and penalty clauses. The price of the service is a major consideration in some localities and responding companies have offered either a low price tier of service or discount rate.

Privacy and Leased Access

The right of subscriber privacy and the ability for entrepreneurs to gain access to cable systems are major questions grappled with by municipal officials. Regarding subscriber privacy, some of the pertinent questions asked by municipalities are: what types of information will be collected and how long will it be stored; who will have access to the information; if there are errors in the information, under what conditions will they be corrected; and how is a dispute between the company and the subscriber to be handled. As more sophisticated cable systems are built and more transactional and service offerings are provided, the amount of personal information that can be gathered will increase significantly. An in-depth discussion of subscriber privacy can be found in Chapter 9 of this book.

The ability of local, regional, and national entrepreneurs to gain access to cable systems is a hotly contested issue. Previously discussed ownership trends, pending deregulation legislation, and contentions of the cable industry that they should enjoy the same First Amendment rights as newspapers underscore the industry's objection to providing leased channels. Municipalities, on the other hand, contend that subscribers have a right to select from diverse information sources and that entrepreneurs other than the cable operator should have access to a virtual monopoly which utilizes the public's rights of way. This complex issue is addressed in Chapter 10.

Community Access

One of the decisionally significant factors in franchising for the last several years has been company committment to community access for the municipal, educational, and public sectors. Municipalities have scrutinized the number of channels, equipment packages, and financial support offered. A growing trend, based on successful access operations, is the establishment of a non-profit access organization to provide the equipment, training, and support personnel for access. The cable company and the city provide the funding through a portion of the franchise fee for the non-profit organization. Well-funded community access centers have found that local residents do indeed produce and watch local programs. The provision of community access and financial support is also an area for municipal consideration during refranchising and renewal. The background and current status of community access is discussed in Chapter 6.

Interconnection

Interconnection of cable systems on a local, state, and national level has increased dramatically in the last four years. New statewide interconnects have been initiated in New Jersey and Pennsylvania, regional sports networks are being formed in a number of major markets, and on the local level, nearby jurisdictions have been sharing access programming. Since interconnection provisions in older franchises usually are non-existent or unenforcable, municipalities see refranchising and renewals as an opportunity to address local interconnection. Issues that need to be addressed in new franchise agreements are: which systems will be interconnected free for public sector use; what technology(s) will be utilized (e.g., hard-wire, microwave); what the time table for implementation will be; who will pay for the system; and who determines eligibility to utilize and schedule public sector channels.

Institutional Networks

Extensive inter-institutional use of cable television for data transmission, energy management, training, and teleconferencing is envisioned for the major urban markets. In these markets separate cables running between public institutions and businesses form an institutional network that will serve the commercial and public sector markets. The need for inter-institutional communications should be considered during refranchising and renewal. If such a need exists (e.g., fire station personnel training, school building energy management), municipalities can determine if it is economically feasible in the particular city to provide for inter-institutional communication through a separate channel or through scrambled channels on the subscriber network. The channel capacity available, the cost of use, if any, the buildings passed, and the number of drops provided for this purpose are all elements to be considered by the municipality in determining if a cable system will meet present and future institutional needs. Suggested Request for Proposal (RFP) language for institutional network and community access is provided in the Appendix.

Franchise Fee

Many of the franchises due for expiration have franchise fees of one to two percent of net company revenues. Federal Communications Commission rules in 1982 allow municipalities to request a franchise fee of three percent of gross revenues with the ability to gain a waiver to charge a five percent fee for use in funding regulation and local programming. The amount of the franchise fee and the revenues it is derived from are substantive questions for a municipality at renewal and renegotiation time.

Overall Service Package

The overall technical, consumer service, and local programming package is the overriding consideration for renewal and refranchising. Municipal expectations are discussed in Chapter 4 where a case study of the Dubuque, Iowa, renewal process is described. The capacity of the system, number of services offered, funding for access, and other items mentioned throughout this book, should be major considerations in deciding on what renewal or refranchising procedures are utilized and in making the final determination on the provision of a new franchise document.

Use of Consultants

Municipalities increasingly utilize consultants in the refranchising and renewal processes. Consultants can provide a nationwide perspective on the industry as a whole. They also offer a multidisciplinary approach which can provide: advice on local, state, and and antitrust laws; a sound foundation for franchise review; an assessment of which procedures to follow; an assessment of community needs; an evaluation of the quality of technical and service proposals; an evaluation of the financial viability of the system; and advice and guidance to city staff, council, and citizen advisory boards on other major and minor municipal considerations. Cities utilizing consultants find it most beneficial if they are brought in at the initiation of the franchise/renewal process. Often bids from potential consultants are solicited and selection is based upon the quality of service that can be provided and the recommendations of other cities they have worked with.

THE RENEWAL AND REFRANCHISING PROCESS

Initial Steps

The first question municipal officials ask is when should a community begin formulating the timetable for renewal or refranchising. The answer is simple—start early. Some cities have waited until six months prior to expiration and found themselves ill informed and ill prepared for the complexity of modern cable refranchising or renewal. Other cities have been requested by operators to consider beginning the process as early as five years prior so they can refinance, rebuild, or both. In some cases, companies have rebuilt in hopes of enhancing their position with the community when the franchise expires. Cities have found that initiating the processes two to three years in advance provides the necessary timetable to gain the proper results.

There are several preliminary steps a community can undertake when initiating the renewal or refranchising process. Following is a brief description.

Establishment of a Cable Advisory Committee: If there is no existing Cable Committee or Board, many cities establish one which includes representatives from a cross section of the community (e.g., library, education, civic organizations, subscribers). Sometimes one or more council persons also serve on the Committee. The role of the Committee is usually described by resolution of the city council and members are appointed in the typical manner for each respective city. The selection of the Committee is extremely important since its members will fulfill a time-consuming and important function—to seek information regarding cable and advise the council on how to acquire the best possible cable service for local residents.

Information Gathering on Cable: The first important step for the Cable Committee and Council is gathering information on the state of the cable industry, current technological innovations, community communications, and the status of area franchises. This information will provide the basis for decisionmaking throughout the process.

Community Needs Assessment: In order to determine the present and future needs of the community for internal and regional communications, a needs assessment could be conducted. This would consist of public education seminars, individual and organizational questionaires, in-depth interviews, and an analysis of how community needs relate to the design of the cable system and the financing and implementation plans for local programming.

Review and Update of Cable Ordinance: The existing cable ordinance and/or a franchise should be reviewed and updated. The sophistication of regulatory documents has increased dramatically as urban areas have franchised. Additionally, traditional problems with industry promise vs. performance have led to many new regulatory provisions. For example, slow or no construction in low income areas was responded to through ordinance provisions which allow for the levying of financial penalties and in some cases possible license revocation. The importance of revising the ordinance cannot be underscored sufficiently. It is the centerpiece of contract regulation and compliance.

Evaluation of Company Performance: An evaluation of company performance is often the first step in renegotiations. The evaluation encompasses technical performance, ordinance compliance, quality of service offerings, overall financial standing, customer relations, financial support for local programming, and responsiveness to municipal and community requests. This is sometimes done prior to the decision of which procedure to follow.

Analysis of Economic Potential: An analysis of the economic potential of the system would provide the Committee with a basis for determining the acceptable system design and service offerings. Often, a market study of consumer willingness to pay for new services is called for. This is because the expandability of service tier purchasing varies from community to community.

Consideration of Multi-jurisdictional Cooperation: Each city, after discussions with municipalities embarking on similar cable processes, should decide whether multi-jurisdictional cooperation, and the aggregation of subscriber base it implies, is possible and would serve the best interest of the community. If multi-jurisdictional action is contemplated, it is useful for all participating municipalities, depending on state law requirements, to agree at an early stage, through municipal resolution, to participate in the process. Usually, municipalities adopt the same ordinances, go through the various processes together, and award similar franchises.

Public Participation: Public input on service, programming, and process is an essential element in renewal and refranchising. Normally, public participation is solicited in a variety of ways throughout the process (e.g., informational meetings, public hearings on the franchise and offerings, and written comments on offerings and consultant's reports).

Procedural Options Paper: After the previous steps have been taken, a procedural options paper can be prepared which will give the committee and council the legal and procedural pros and cons of the renewal, refranchising, renegotiation, and alternative ownership options. This will assist the decisionmaking on which way to proceed. (A description of the legal considerations involved can be found in Chapter 3.) The Council usually makes the final decision on the approach to be taken.

Renewal Procedures—Automatic

Automatic renewals usually include clauses which provide for compulsory renewal or renewal at the option of the cable operator. In these situations only a few procedures are necessary or legally permitted.

Legal Analysis: A legal analysis of the franchise will determine if indeed the renewal is automatic in nature or if there is sufficient leeway for negotiations to take place.

Performance Review: Some automatic renewal clauses call for the system to be in substantial compliance with the franchise. In these situations a thorough review is called for. If the system is in compliance, the municipality would proceed with the renewal. If not, the legal analysis would direct the next steps to be taken.

Municipal Action: Assuming an automatic renewal requirement, the city, by municipal ordinance, renews the franchise for the specified length of time.

Renewal Procedures—Renegotiation

Renegotiations are handled differently by each community. These differences can in part be attributed to system capability and offerings, quality of customer service, and the relationship between the company and the city. Following is a suggested method of proceeding in a renegotiation setting.

Preparation of Specifications: Similar to development of an RFP, specifications are developed to determine what technical and service offerings the incumbent company would offer. The evaluation criteria to be used will also be described in the specifications. By preparing specifications, a municipality is establishing the framework for negotiations and can solicit information on the many aspects of cable service. It also provides a basis for comparing the service offerings to those offered by other companies competing in franchise and renewal processes in similar-sized cities. While comparisons are mitigated by many factors (e.g., density, amount of underground construction), they do assist cities in determining the appropriateness of the company offer.

Proposal Evaluation: The company responds to the specifications in the form of a proposal. The city reviews the proposal, posits any questions it might have, and, after it receives the answers, evaluates the proposal and determines if it meets community needs and standards.

Public Participation: Public participation in renegotiation varies from very little to a series of hearings and written comments. Cities that have adopted public participation procedures have found them very useful.

Negotiation: Following the proposal evaluation, negotiations take place. During negotiations, the city and the company come to final agreements on offerings, construction, and regulation. Once negotiations are completed, a franchise is awarded and signed.

Refranchising Prodecures—Competitive Bidding

Refranchising procedures are basically the same as those undertaken in granting an initial franchise. There is some variation in procedure based on state law or regulation. For example, in Iowa, municipalities must have their choices for the franchise affirmed by a referendum, and in New York, municipalities must comply with minimum state standards. Following is a brief description of procedures commonly adopoted.

Preparation of a Request for Proposals (RFP): The municipality normally develops an RFP on which all potential applicants provide pertinent information on experience, financial capability, system design, rates and financial plan, local programming, and entertainment and non-entertainment service offerings. The RFP is designed to provide the basis for comparing the capability and offerings of prospective applicants. Normally, the basis for evaluation and the highlighting of any criteria especially important to the municipality are provided so that the applicants are aware of community priorities (e.g., senior citizen discount, hispanic program service, local programming). Announcements of the RFP are usually published in local, state, and national publications, and interested parties request copies of the RFP. Thirty to fifty days are usually given for applicants to prepare their proposals.

Proposal Evaluation: The proposals, once received, are evaluated based on the criteria established in the RFP. A two-step evaluation process is common, with preliminary and final evaluation stages. The preliminary evaluation is that—preliminary, with questions on unclear parts being asked of applicants. Responding applicants are sometimes asked to critique the preliminary evaluation. Responses are considered and a final evaluation report is prepared. This provides the basis for selection of the applicant. Sometimes applicants will be extremely close in experience, capabilities, and offerings and decisionally significant areas will have to be determined (e.g., a city may select a company offering cheaper rural service over a company offering a longer institutional loop).

Public Participation: Public hearings and the submission of written testimony takes place at various stages including hearings on the RFP, the preliminary evaluation of the companies' proposals, and the final evaluation of applicants.

Franchise Negotiation: Once a top applicant has been selected, negotiations are conducted to "dot the i" on offerings and contract agreements. When negotiations are completed, the franchise is awarded and signed.

Refranchising Procedures—Alternative Ownership

There has been an increase in the number of municipalities considering alternatives to private ownership. Valpraiso, Florida, for example, established a municipally-operated cable system because it was unhappy with the performance of its existing company. In approaching an alternative ownership form—municipal, cooperative, joint venture—a city can proceed, depending on state or local law, through a refranchising or alternative ownership procedure. Following are alternative ownership procedural steps that parallel those undertaken in initial franchising procedures.

Preparations for City Decision on Alternative Ownership: The preparation for a City decision in this area varies by locale but includes one or more of the following: an analysis of past company performance, a fifteen-year financial plan, a study of legal and structural options, and an analysis of financing possibilities.

Public Participation: Public participation is solicited throughout the alternative ownership process as it is in renegotiation and refranchising. Common points for public hearings, meetings, and comments are initially for educational seminars, hearings on whether to go an alternative route, and hearings on the proposed service offerings. As with other procedures, state law needs to be considered (e.g., if a referendum is required).

Preparation of Specifications: The city may prepare specifications similar to those prepared in the renegotiation process and request that the co-op or public utility or city staff or newly created entity prepare a description of the service offerings, system capacity, etc.

Evaluation of Offerings: Once the entity designated to respond to the specifications has done so, the city evaluates the offerings and asks any questions.

Final Proposal Development: The final proposal is developed after the evaluation is provided and any additions accommodated. Once the final proposal is acceptable, the city awards the franchise.

CONCLUSION

Cable as an industry has evolved dramatically—technologically, programmatically, and operationally—since its inception. Refranchising/renewal provides municipalities with the opportunity to see the local cable system benefit from this evolution. Through careful study and adherence to successfully utilized procedures, a municipality can ensure that the newly licensed cable system will meet the telecommunications needs of local residents now and in the future.

Whereas this chapter has provided an overview of renewals and refranchising, other chapters provide in-depth review of several critical areas: how programming and technical advances affect the local system; what legal considerations municipalities need to be cognizant of; what municipal and oeprator expectations are; how community access and rural service can be provided; how privacy and leased access issues can be grappled with; and what RFP language is needed in the often neglected area of community access, institutional networks, and future services.

Thomas A. Muth

2. Cable Television Programming and Technology

BACKGROUND

Programming and technology in present cable television systems differs greatly from that offered when the currently expiring ten to fifteen year franchises for cable service were granted. Between 1967 and 1972 cable service typically consisted of a twelve-channel maximum selection of one-way (downstream) entertainment programming from within a zone of a hundred or so miles. Imported or microwave-relayed signals from more distant broadcast stations and locally generated "billboards" featuring time, temperature, and barometric readings occupied the balance of channel line-ups.

Systems operating and proposed in the late sixties and early seventies differed technically according to community size, geographic location, community leader sophistication, and operator interest. Introduction of a system in Akron, Ohio, for example, with sixty-some video channel capacity was the talk of the cable industry. Cypress Cable (later to become Warner and then Warner Amex Cable), the Akron franchisee was hard-pressed to economically justify its investment in the multi-cable "mega-bandwidth"

Thomas A. Muth is Associate Professor of Telecommunications, Michigan State University, and a principal and co-founder of the ELRA Group, a telecommunication research development and consulting firm with offices in East Lansing, Denver, and San Francisco.

system. Two-way and interactive cable television were discussed, but available only in promise and speculation. The generation of alphanumeric video screens was expensive and rarely found. Portable and remote production was beginning, but always monochrome. Transfering a stored video signal from one tape format to the then-recent ¾ inch helical scan videocasette format, was necessary. The transfer was generally possible only through expensive time base correction, usually found only at the larger broadcast centers. Domestic communication satellite service was promised. Satellite delivered programming and pay cable services were in planning.

At this time cable television had fallen from its initial high revenue/low cost development status. FCC restrictions on distant signal importation and use of syndicated programming and requirements for carriage of local and significantly viewed broadcast signals disadvantaged the cable industry. Uncertainty prevailed as the FCC plodded through seemingly endless cable rulemaking processes. Cable operators and communities granting franchises during this period were sincerely in doubt as to the long-term prospects for the industry. Then dramatically, the FCC announced its "open-skies" domestic satellite policy. Almost concurrently it released new cable rules. These moderately relaxed the signal importation restrictions. Shortly thereafter pay cable services began to succeed. As a result of the new cable rules, cable operators and their financial backers could justify risking development in communities with existing UHF and VHF television broadcast stations. The coalition of satellite delivered pay cable services and moderate relaxation of the cable importation restrictions (along with improved general business conditions) proved sufficient to revive the cable industry. Cable stock prices rose and capital for expansion became more available by the mid to late 1970s.

In considering franchise renewal and redevelopment it is necessary to recall conditions prevailing when the prior franchise was granted. This consideration requires more than assessment of the technology or programming of ten to fifteen years past. It calls for a careful recollection of promises and potentials then made and discussed. Reexamination of the familiar "blue-sky" promise of the "wired city" of the late sixties and early seventies is indicated. Today's present technology may fulfill some of the promises, desires, needs, and demands that remain unfulfilled in communities. The upgrading, remarketing, reevaluating, and retrofitting of systems and services at franchise renewal and in renegotiations suggests enhanced opportunities for revenues and realization of new business for cable operators. Future history should recall the mid and late 1980s and early 1990s as a time of revitalization for cable systems in small and mid-sized U.S. communities. If this general objective is realized, the dream of the wired city, nation, and global-village world will be close at hand. This realization demands wisely informed, long-range structural planning to enhance technology and programming services and integrate them with the needs of the regional, state, national, and international society. Integration of social and economic objectives through structural planning and policy making *can* move the formation of a broadly based information valued society. The absence of such planning implies a resurgence of analogies to the chaos that hindered early voice-grade radio-telephony and broadcasting.

The analysis in this chapter looks to the lessons of history, in this case the specific history of programming and technology in cable television, as a grounding and basis for casting the future. The intents and purposes of early cable programming and technology, their actual uses (or their lack of use) will instruct on what the renewed, retrofitted, refranchised cable systems portend.

Refranchised, rebuilt, renegotiated, retrofitted, or renewed cable systems and services differ markedly from new systems or "builds." In designing cable services for uncabled communities, novel alternative services may be offered and if well-marketed, may succeed. In renewals, or systems or service upgrades, wholly new service offers are risky. The existing cable subscriber has developed use patterns and preferences. These are likely ingrained and must be discovered and defined. This implies that the cable operator has a deft sense of the existing and potential market for the local service. It implies quintessential perception of community wants and demands. The operator must be able to distinguish what services the community, its leaders, and citizens want from what they will actually demand, subscribe to, and purchase.

In this context, the community must be understood as a market and the cable operator (and possibly the local franchising authority) must undertake careful market analysis and evaluation. This frequently indicates the need for research to establish present suscriber satisfaction, information on satisfied and unsatisfied subscribers, and subscriber and non-subscriber interest in new types of cable programming and related services. Research indicates unexpected national findings that require careful analysis for local levels when revising cable services. For example, in general, subscribers tend to be younger, less educated, more affluent, and more likely to be married and have children than non-subscribers. Naturally this generalization, based on valid national research, is subject to considerable variance in local application. Examples of affluent, highly-educated communities demanding expanded programming and related services exist. The factors that render one community above or below the norms are revealed only through careful local analysis of cable service use and behavior. Comprehending this complex mix of contemporary social, individual, and other components calls for professional market research.

Knowledge of existing patterns of use of cable service is tantamount to redevelopment of the system or the services such as sought in renewals. The history of viewing patterns, program preferences, and acceptance of innovations all must be fed into plans and proposals for redevelopment or upgraded cable systems and services. Inadequate analysis of programming and related services has led to both excessive *and* inadequate service and technical offerings in renewal settings. Programming and service evaluation precede technical reconfiguration. Need and demand for services should lead technical designs to avoid excessive or inferior human and economic investment and capitalization in renewal.

CABLE TELEVISION PROGRAMMING

Cable television programming has several aspects which reflect unique historical development and technical characteristics. These must be carefully reviewed upon renewal. They must be compared with contemporary programming considerations such as tiers or groups of programs and regional programming.

Background

Cable programming had early false starts as scattered cable operators attempted local origination of advertising supported programs. Simple game programs, such as bingo, were perhaps the most successful of these early efforts. Other operators attempted creation of daily or weekly news programs. But most early local commercial cable attempts failed for lack of production organization and lack of audience.

Production organization was difficult because cable operators usually had little internal production management experience. The "local origination" programs were frequently produced by student interns from colleges and high schools.

The early lack of production depth could perhaps have been dealt with if audiences would have been developed for commercially sponsored locally produced material. However, many cable system operators attempted to imitate broadcast programs in techniques and style with local advertising supported origination. Commercially sponsored local cable news and sports imitated, albeit poorly, network and broadcast television programs. The local "commercial" efforts paled as viewers switched from expensive full color local or national broadcast quality television production, to low contrast monochrome cable local news and public information programs. The early attempts at commercial local origination were not to succeed.

The failure of this specific genre of production was probably fueled by the FCC's so called "3500 Rule." This policy, adopted in the early 1970s, required that cable systems with 3500 or more subscribers have facilities for and offer local original production. Though adopted, this policy was never enforced by the FCC. It did, however, become the basis for precedential judicial law when the Midwest Video Co. challenged the rule. In a 1973 decision the U.S. Supreme Court reversed an appellate position that the FCC was requiring cable operators to enter the local broadcast-type origination service without authority in commercial operation of cable systems to require that they originate local programming. The court found the local origination with its implied potential for advertising revenues sufficiently similar or "ancillary" to broadcasting to allow the FCC to require programming origination. As noted, the FCC never gave efficacy to this policy and even with a supporting Supreme Court decision, it abandoned it altogether.

The practical, administrative, and judicial history of the "3500" local origination rule offers a handy and well-documented distinction for use in revising service programming. The FCC rule had mandated the cable systems having 3500 or more subscribers to originate local programs. At the time this rule was promulgated, a number of towns and villages were either adopting new cable franchises or in a renewal process. The notion that an access to "media" had gained favor through a theory that insisted a right to access to the means of expression was implied in the First Amendment. Local franchising authorities often found merit with this position. As a result many local cable franchises and ordinances required that the cable system provide access to facilities for program origination to members of the community.

These requirements were usually broad and implementation was left to practice. In some communities telephone-like booths with fixed camera and microphones were installed so that members of the general public could "walk off Main Street" and transmit an audio-visual message to the cable subscribing public. Usually this approach failed. Few used the service. The fixed camera and microphone position inhibited message and information development. The sporadic nature of program schedules made audience presence uncertain.

Videotape technology allowed potential storage and frequent replay of this brand of program; however, few cable systems possessed the needed recording equipment. The approach was intended for low cost live message delivery. Even systems with recording capacity hesitated to record. The cost of tape and difficult economics of the cable industry found little cooperation with this "walk-up" transmission approach.

In 1973 the FCC's exercise in "creative federalism" established the FCC rules. These rules required that cable systems in the largest fifty markets provide facilities for "government," "educational," "public," and "leased" access facilities. A careful long-range comprehensive federal policy is generally not found in these rules. However, while no long range policy objectives are clear, by the *fact* of adopting these requirements the FCC left a mark on U.S. telecommunications. "Access" dedicated services became a function of cable television and produced lasting change.

In 1979 a case was brought by Midwest Video Co., which asserted that public access communication requirements were not within the authority of the FCC, and which was sustained by the U.S. Supreme Court. This decision left the legal and national policy status of public access and other access cable services in doubt. However, in the duration since adoption of the FCC Cable Rules, local governments and citizens accepted and posited further use for access communication. As a result most cable operators support access communication in existing, new, and renewed franchise and license settings.

As a result, in systems with larger subscriber bases, generally those with over 3,500 subscribers, some form of cable access has become a tradition. In reviewing systems of over 3,500 subscribers access communication facilities and facilitators are generally included. Where systems are substantially larger, the size of the facility, number of access channels, studios, support equipment and mobile and remote origination service is increased. A system passing 30,000 residents should contemplate providing more than one studio site and related equipment. Review of recent large community franchise and refranchise development suggests that about 10 percent of the gross capital of cable systems is being invested in government, education, and public access equipment and personnel to facilitate development.

Stories of unexpected success with local access to small format video production facilities abound. As local channels offered opportunities for subscriber-wide distribution in the form of local video, film and live production, interested citizens were attracted to use.

PRESENT DEVELOPMENT

Today cable programming assumes several forms at local levels. Access communication has become well accepted by local amateur, semi-professional, and institutional producers. Government, education, and public access cable television productions and production units are found in well over a hundred locations. In addition, as larger metropolitan markets are developing cable, with system and community interconnects, interest in local commercial origination production is being rekindled. Each form should be included in renewing a franchise or upgrading a system. The amound of access service should increase in proportion to system size.

A variety of access communication services can serve as models for a renewed cable system. Schools use alpha-numeric character-generated video typography to inform communities of school closings, lunch menus, special programs, etc. City Council, government committees, and related meetings have become highly popular in many communities. Amateur video producers avail themselves of equipment to learn use of production technique, recalling the time honored contribution of the early "amateur" radio broadcasters. Many of those amateurs engendered the wholly new concept of electronic broadcasting. Similarly, many public access producers have been and are purposely tinkering with alternative media and in the process of engendering new communication forms.

National Production in Cable Television

Numerous national programming services have been forming to offer their wares to the cable television industry. These generally fall into the following categories: (1) distant broadcast stations either imported by microwave or satellite and (2) original productions produced expressly for distribution by cable television. Each must be considered in revising cable program offerings at renewal. The latter programs can be put into three classes: (1) advertising supported productions, (2) programs supported within the frame of basic cable subscriber services, and (3) pay cable programming. The pay programming may be further divided into periodic pay programming and pay-per-view programming. It should be noted that very few systems currently operating can monitor pay-per-view programming; however, this technology is rapidly developing and must be considered in all renewal franchise settings.

Distant imported broadcasting station programming is exemplified by the "superstations" such as Atlanta's WTBS or Chicago's WGN. Programs produced for basic cable service include special interest programs such as Nickelodeon for children, Christian Broadcasting Network (CBN), Entertainment and Sports Programming Network (ESPN), and Home Theatre Network (HTN). Pay services include Home Box Office (HBO), The Movie Channel, and Cinemax. All of these production services are programmed nationally and rely on national distribution by satellite. They must be carefully distinguished from local production. They are more closely identified with the character of standard broadcasting services directed toward mass audiences. The degree of specialization found in the nationally distributed productions is directed toward interest commonality rather than geographic location.

Program Tiering

Tiering is the grouping or packaging of program services into varied classes or types of service. Tiering allows combinations of pay cable, basic cable, advertising supported, and other newly developing services.

The rate charged for individual tiers may differ. The classic cable service providing the transmission of local broadcasting and local origination cable services often becomes a "basic tier" in a franchise renewal. (In some renewals, cities are requesting a form of "universal service" which provides free or low cost service for all subscribers.) Pay cable services may be grouped together to form yet another tier. In larger capacity systems several pay services may be offered and marketed successfully.

The renewal is an opportune moment to gradually develop policy for restructuring programming line-ups. Basic service may be mixed with pay service to form more appealing packages. The basic service may include local origination and access and thereby enhance their positioning for the community.

Also, satellite delivered special or general purpose programs such as USA Network and sports may be mixed with local, basic, and pay service to create attractive tiers of varied programming. The combination of programs that can be offered is limited only by the imaginative resourcefulness and research of the cable system and community leaders. Each individual community is more or less amenable to different program combinations and to multiple tiers. Larger communities, say with 15,000 households passed, are amenable to greater variety and complexity of tiered programming. Smaller communities, those of the 3,500 or smaller household variety, present more difficulty for multi-tiered varied program packages. The number of tiers and the combinations of local satellite, pay, and non-pay programs must take the community size, absolute spendable income in the market, and present viewing or use patterns into account. New tiers, combinations, and program offering should be added incrementally in a refranchising setting. This will best accommodate existing viewing and use patterns by the community, maintain subscribership, and reduce the risk of unanticipated "churn," or subscriber cancellation. Abrupt drastic program line-up changes, wholly restructured program tiers, may impair cable subscriber user behavior unless carefully considered by both the cable system operator and community leaders.

Regional Productions

The last area that should be considered in programming for cable television is the very new area of regional programming. These programs are only now developing and are principally based on regional interest in sports and sporting events. There is some additional interest in regionalization of news effort. Finally, education production is developing within and between some states that concentrates on vocational information and community college information. In most of the educational cases, the productions tend to feature lifelong or continuing education formats. Again, these program types should be considered in cable service revisions. They may be integrated with tiers and programs developed for franchise renewals.

Cable television programming required significant time for initial development; however, it is now fairly clear that techniques and types of production and delivery of programming is developing at greatly increased rates. Potential in the cable field is vast and differentiated. Opportunities exist in wide ranging areas of communication to develop programs and services that respond to the demand potential associated with the new multi-channel metropolitan cable systems. Economic models that seem best to relate to cable television productions are those derived from the book and magazine publishing industry. This suggest diverse programming in style, quality, price, and profit. Larger communities can demand and expect more program variety than small communities; it is safe to expect, however, that *all* renewed cable franchises should reconfigure, increase, and diversify a program and service mix.

CABLE TELEVISION TECHNOLOGY

A discussion of cable technology should begin with an overview description of a typical cable system. To assure clarity, it is best to trace cable technology from early system development and add new techniques and equipment as they appeared in history. From this approach it is fairly easy to infer technological considerations for redeveloped cable service.

Background

The early systems usually provided no more than six channels of television and many offered only one or two channels. Some of the six-channel systems had excess capacity because there were few television stations broadcasting within the range of the relatively inefficient reception antennae. Many of these systems adopted a practice of training a camera on a clock or newswire service. Later, a small camera was installed to rotate from a clock to a weather instrument array and then to a third panel where the camera might pick up hand-written community messages.

Where cameras were in use, some systems began to provide audio and video public announcements. In some rural communities, as previously noted, bingo games were transmitted and became quite popular with local merchants providing prizes of merchandise. Some of these small communities evolved sporadic, local programs. Infrequently the cable system office or headend had space for a studio. Early cable studios were usually no more than a drape behind a fixed camera and microphone. Few individuals used the facility and there was usually no formal program provided either by the cable system operator or local community group. This type of service is unlikely to find success in a renewed franchise.

System or cable plant technology started with parallel pair antenna wire but quickly thereafter the newly introduced coaxial cable, that was being used for transcontinental television transmission, began to find its way into the CATV systems. Virtually all of the multiple-channel cable systems quickly introduced coaxial cable. The principal advantage of the coaxial plant was that of expanded bandwidth. The coaxial cable was rapidly undergoing technological development so that greater and greater bandwidth could be transmitted with virtually identical supporting utility poles and supporting plant. By the time that most of the smallest U.S. communities had cable, service equaling twelve channels had become common. It should be noted that each video signal requires about six megahertz of bandwidth; some bandwidth is also needed between channels to inhibit co-channel or cross-channel noise and interference. Thus a twelve-channel cable system has about 72 megahertz bandwidth. Virtually all of this capacity was originally devoted to video broadcast signal retransmission. The cable operators received a broadcast signal, processed it into a channel transmitter, and distributed the signal on the system. The signal was boosted at the headend, or central receiving and distributing site, to provide substantially improved reception in the community.

Power for the carrier signal transmitted on the early cable system was introduced at the headend and boosted by line amplifiers and power supplies throughout the system. It was unlikely that the twelve-channel cable signals would withstand transmission beyond about five miles from the headend, and most operators found they had to limit the cascade, or number of line amplifiers, to about fifteen to maintain acceptable signal integrity. When signals were transmitted beyond this distance and cascade, the interference, noise, and general degradation became unacceptably pronounced and subscribers often cancelled the cable service. In a refranchise setting new technology will reduce noice and interference in transmission and the cable plant can be constructed with larger trunk runs and therefore greater system size.

In a slightly later phase of development, cable operators began to capture broadcast signals in distant cities and transmit them to a remote location by microwave carrier. This became known as importation of distant signals. In this manner, stations broadcasting at distances of several hundred miles from the location of the cable system could be received by the system's subscribers. Where common carriers were employed to provide the importation carriage, the courts held that the FCC had jurisdiction over the cable systems carrying the signals. In a later case the courts held that the general operation of cable television systems was subject to the control of the FCC as a species of broadcasting whether signals were imported or not.

As the number of signals available for transmission by cable television systems increased through better techniques such as antenna design, signal importation, and better local alpha numeric information, the twelve position click-stop VHF tuner, typical of most television receivers, was exceeded. To provide 24 channels of 6 megahertz television, many cable operators increased system capacity to 20 or 24 channels and introduced "A-B" switches which allowed the subscriber to select signals of an additional 12 channels for viewing. The advent of the A-B switch was shortly followed by the multi-channel set top converter, which processed 24 or 36 signals and allowed the cable system operator to place signals on channels different than those of original broadcast. This permitted elimination of different forms of interference from adjacent channels. It also allowed the cable operator to expand the channel offering substantially.

Channel Capacity

Twelve-channel systems are still found in about half of the franchise areas in the country. These systems are older and in many cases their franchises are terminating or may have been renewed within the last year or so. The twelve channel system is usually found in communities with less than 15,000 households. Communities of more than 15,000 households and franchised in the early to mid 1970s often have 20 or 35 channel systems. Their franchises usually have a 10 to 15 year duration and will be expiring in the mid to late 1980s. It is not too early to begin to review the service delivered by systems of this capacity.

Modern residential cable service technology should be determined by market analysis. Generally such analysis has indicated that system bandwidth for residential subscribers depends on issues such as community size, size of households, proximity of the community to large multi broadcast cities, and local propensity to consume cable service. Thus some new systems offer residential subscribers over 100 channels, some 50 or 52, and others satisfy community demand with 26 or 35 channels. The question of bandwidth for residential subscribers is thus more of a marketing than a technical question.

Franchises granted since the 1970s are generally found in larger communities, including major suburbs in metropolitan areas. They frequently offer 50 plus channel capacity. The largest capacity systems offer over 100 channels of residential service with independent or integrated institutional network and interconnective service. While systems of this capacity will generally not be the subject of renewal for 10 or 15 years hence, close coordination between the system operator and community leaders is highly adviseable. They should determine if the cable system goals, market plans, and service directions coincide with plans for general development by the community. A regular systematic process of evaluation and even interim renegotiation of services can benefit the cable operator as well as the community.

Using interactive and two-way cable and related innovations developed by researchers, the cable television industry initiated two-way interactive cable television in the late 1970s. Coaxial Communications and Warner Cable installed systems at Columbus, Ohio. Two way interactive cable has been the subject of much discussion and of little commercial success. The services have long range implications for alarms, monitoring, remote signaling, and pay-per-view programs. Despite the present question of the success of these services, virtually all new revisions and renewals of cable service technology should carefully consider two-way cable in renewals. Two-way interactive cable has obvious long-term future application. Care should be taken in adopting standards for two-way interactive service to assure the franchiser and franchisee agree. Confusion often focuses on whether a system is two-way *capable* or two-way *active.* The "capable" system is one which requires the addition of return amplifiers or other equipment for operation. An "active" two-way system can presently transmit up and downstream service.

Expanded Networks

Another technical facet of cable service that has developed recently is the expanded networks. Today's cable systems generally offer service to residents, service to private and public institutions, and interconnection service to and from other cable systems. The technical design for this portion of service frequently calls for transmission on a separate trunk cable. Many designs however successfully integrate institutional service with residential service. The proximate or ascertainable demand for institutional services controls the technical design. This is again a market question. If demand abounds among private or public institutional users, a separate cable plant may be warranted. If market analysis indicates a relatively small number of business or public potential users, the institutional network is more likely to integrate with residential service. Naturally, provisions for expansion and construction of separate trunk and plant should be treated in any renewal or upgrade consideration.

A major new technical component of cable service is interconnection. This is a hybrid of policy and technology. The renewing parties will have to develop a plan and policy as to interconnection of contiguous, non-contiguous, and other area cable systems. Given development of such a plan, a policy comprehending an area-wide mutually agreeable system design is needed. In many circumstances interconnection initiates a second order or layered cable system. This new system can be separately franchised and operate to serve as a "cable system's cable system."

SUMMARY

This discussion should suggest the degree to which programming and technology in cable redevelopment are mutually dependent. It also should suggest the degree to which careful analysis built on well developed market researching, planning, and policy are required for renewals.

David M. Rice

3. Legal Considerations in the Refranchising Process

At every stage of a city's refranchising process legal issues arise and must be addressed by the city, the incumbent cable system operator, and by any other companies that have an interest in being franchised by the city. Discussion of these issues must, however, also touch upon technological, financial, and other issues with which the "legal" issues are interwoven.

While most cable operators are well informed about the legal aspects of cable franchising, the cities with which they deal are generally far less knowledgeable. Cable operators usually have both in-house counsel and the services of outside communications counsel, but few city attorneys are expert in this area. As an increasing number of cable franchises are reaching the end of their terms, a similar knowledge gap is developing with respect to the legal aspects of refranchising. Even those cities which have closely monitored their cable systems and maintained an active enforcement program may be ill equipped to cope with many of the legal problems that may arise in the course of refranchising.

In addition to the numerous problems peculiar to the refranchising process, most of the legal issues that normally arise in initial franchising situations are likely to recur in connection with the award of a franchise which will renew, replace, or supplement an existing one. A city engaged in

---

David M. Rice is currently Associate Professor of Law and Associate Director of the Communications Media Center, New York Law School. He is also a legal consultant to cities on cable franchising and refranchising.

refranchising thus will have dealt with those issues before. Nevertheless, normal personnel turnover and intervening changes in the law may make it necessary to provide education or reeducation in these areas to the members of the city's legal staff with responsibility for cable refranchising.

COMMENCING THE REFRANCHISING PROCESS

When and How to Begin

One of the most common errors made by cities is that of waiting too long to begin the refranchising process. Often a city will wait for the incumbent franchisee or a would-be competitor to take the initiative, or will do nothing until a matter of months before its current franchise is scheduled to expire. As will be discussed below, such delay may impair significantly the city's legal position and limit its options for the future. In cable refranchising, as in other areas, "preventive lawyering" is sound practice. Preparations for refranchising should thus be commenced long before the existing franchise expires; even two or three years is not too long in advance to begin planning.

A thorough evaluation of a city's cable system thus requires the involvement of persons with expertise in law, engineering, and finance, as well as in cable programming at both the national and local levels. Few cities have such an array of experts on their payrolls. A city which has made oversight of its cable system an ongoing priority, however, may be able to amass substantial "in-house" expertise; other cities may have to rely more heavily on outside consultants.

While the extent to which cities have need for cable consultants varies considerably, virtually every city would be well advised to engage a consultant to assist in the refranchising process. A qualified consultant or consulting firm can supply necessary expertise in areas where city staff, including legal staff, are inexperienced; provide up-to-date knowledge of national and regional trends in cable franchising, programming, and system operation; and bring a fresh and unbiased perspective to the city's assessment of its current situation as well as its options for the future. Perhaps most important of all, a consultant can provide invaluable aid in charting and navigating the refranchising process itself; it thus is desirable to bring a consultant in at the very outset of that process.

Reviewing the Existing Franchise

The first task for the city is to study the franchise ordinance or agreement under which cable service is presently being provided. Franchises are generally awarded by a municipal ordinance. In some cities, the terms of the franchise are embodied directly in the ordinance; in others, they are contained in a separate agreement to which the ordinance refers. Since there is little or no legal significance to which of these methods is employed, references in this essay to an "ordinance" or to a "franchise agreement" are generally interchangeable. For convenience, a document containing the terms of a franchise may simply be referred to as a "franchise." [See *Clarification of the Cable Television Rules,* 46 F.C.C. 2d 175, 189 (1974).]

Particular attention should be focused on provisions which deal specifically with the expiration of the franchise or with its extension or renewal. Franchises entered into in the 1960s and 1970s that are now approaching the end of their terms may contain any of a variety of such provisions. These may include a provision for virtually automatic renewal, an option for the operator to renew, or a review procedure for the city to use in deciding whether to renew the franchise. Some franchises are for an indefinite term or presume renewal at the end of the term specified; these naturally make no provision for other contingencies. Other franchises contain elaborate provisions dealing with alternatives to renewal, including post-expiration continuity of service, city buy-out of the system's physical plant, and solicitation of competing bids for future service.

Renewal Provisions

If a city's franchise provides for automatic or near-automatic renewal or grants the cable operator an option to renew, the alternatives available to the city may be far more limited than in other refranchising situations. Renewal on the same terms as are in the current franchise obviously will be less favorable to a city than a new franchise reflecting advances in cable technology and services. Moreover, a franchise entered into a number of years ago may have lacked express provisions covering matters—such as access channels and overall channel capacity—which were then (but are no longer) the subject of FCC rules of standards. [See 47 C.F.R. §76.252 (1979).] A provision in such a franchise which merely mandates compliance with all FCC requirements will, as a result of subsequent FCC deregulation, now have a very different effect than was originally intended; a renewal on the same terms thus could be far less favorable to the city than the original franchise.

Because renewal of the franchise on the existing terms usually would be quite disadvantageous to a city, the city should be alert to the possibility that other franchise provisions may negate or qualify the operator's right to renew. The franchise should be examined carefully to determine whether there are other provisions which make the right of renewal unenforceable. For example, a franchise may condition the right to renew upon an operator's being in full compliance with all of its franchise obligations or with certain specified ones. If, in fact, those obligations have not been met, a city may be able to resist automatic renewal or exercise of an operator's option to renew. Careful scrutiny of an operator's compliance with the franchise is essential to an evaluation of city's rights.

Another possible, but more doubtful, basis for a city to resist enforcement by an operator of a renewal provision or option is noncompliance with the FCC's Franchising Standards, which were in effect between 1972 and 1977. The Franchising Standards included a limitation on the duration of franchise terms to a maximum of 15 years [*Reconsideration of the Cable Television Report and Order,* 36 F.C.C. 2d 326, 365 (1972)]; Commenting on this requirement in 1974, the Commission stated:

> *A franchise calling for a 15-year term with a renewal option at the sole discretion of the franchisee does not comply with the rule. The franchiser must at least review, in a public proceeding, the performance of the system operator and the adequacy of the franchise as well as its consistency with our rules prior to renewal. This is not to say that any bid procedures are required or that any new franchise offering must be made, but simply that a public review of the franchise must be held with the opportunity for citizen input prior to renewal. [*Clarification of the Cable Television Rules, *46 F.C.C. 2d 175, 196-7 (1974).]*

A franchise which was entered into, renewed, or amended significantly during the 1972-1977 period would be in violation of the rule if an automatic renewal provision or an unconditional renewal option would have the effect of exceeding the 15-year maximum term. Moreover, franchises granted and in operation prior to March 31, 1972, which were "grandfathered" and hence not originally subject to the Franchising Standards were required to be brought into compliance no later than March 31, 1977. [*Reconsideration of the Cable Television Report and Order,* 36 F.C.C. 2d 326, 366 (1972).] Effective November 15, 1977, the standards were repealed and reduced to the status of non-mandatory recommendations. [*Report and Order in Docket No. 21001,* 66 F.C.C. 2d 380, 393-94 (1977).] A pre-1972 franchise that was never conformed to the Standards thus also would violate the rule.

A city whose franchise predates November 15, 1977, yet provides that renewal shall be automatic or at the operator's sole option, could argue that the franchise should be deemed amended to conform to the FCC Standards despite the absence of a formal amendatory agreement or ordinance. It would be difficult to read the Franchising Standards as actually effecting the amendment of a franchise, however, since the Commission was careful to avoid that result [*Clarification of the Cable Television Rules,* 46 F.C.C. 2d 175, 197 (1974)]. Instead, it made issuance of the certificate of compliance required for operation of a cable system dependent upon compliance with the Standards [*Cable Television Report and Order,* 36 F.C.C. 2d 143, 207 (1972)]. Therefore it appears that the only sanction for a violation of the Standards would be an FCC-imposed penalty for operating without proper authorization. [See *Reconsideration of the Cable Television Report and Order,* 36 F.C.C. 2d 326, 367 (1974).]

If a franchise contains a clause requiring compliance with all FCC regulations and requirements—a provision contained in most franchises—a city could be content that the failure to amend the franchise to conform to the Franchising Standards constituted a violation of the franchise. As noted above, a franchise violation arguably could bar an operator from exercising a renewal option. However, this line of argument has at least two major weaknesses. First, since compliance with the Standards was required no later than March 31, 1977, the "violation" would have occurred over five years previously, and the operator might successfully argue that the city had waived or otherwise lost its right to object. Second, the city could be regarded as sharing responsibility for the failure to comply with the Standards, since adoption or amendation of a franchise requires bilateral action.

It thus seems doubtful at best that a city could avoid the effect of a renewal provision or option in a franchise merely because of a failure five or more years ago to comply with the Franchising Standards. If, however, a franchise conditions renewal upon an operator's compliance with its franchise obligations, any ongoing violation of those obligations—or at least any serious violation—could provide a valid basis for denying renewal. These conclusions must be regarded as tentative and somewhat speculative, since, to date, they have not been tested judicially. These questions are likely to arise with some frequency as an increasing number of franchises reach the end of their initial terms. Several suits have already been filed against cities that have contended that incumbent franchisees are not entitled to exercise renewal clauses. For example, such suits have been filed in Ventura, California [*Multichannel News,* October 25, 1982, at 17]; and Salisbury, Maryland [*Broadcasting,* August 30, 1981, at 120]. The author is currently serving as a consultant to the City of Salisbury, Maryland.

Review Provisions

As noted above, the FCC's Franchising Standards in effect between 1972 and 1977 required that franchises permit renewal to be granted only after review in a "public proceeding." Consequently many franchises provide for a review of the operator's performance before the franchise may be renewed. The nature of the review and the degree of discretion which a city may exercise in deciding whether or not to renew the franchise will, of course, depend upon the language of the review provision in the franchise. Such a review should be commenced well in advance of the franchise's expiration so that the evaluation of the existing system and franchise can be completed before any of the city's options are foreclosed.

Even if a franchise granted before November 15, 1977, does not require pre-renewal review, a city might take the position that such a review is prerequisite to renewal, on the ground that the Franchising Standards have the effect of incorporating the mandatory review provision into the franchise. For the reasons already discussed, however, this argument would most likely fail. The Standards did not purport to amend franchises directly; they merely required operators and cities to adopt such amendments to existing franchises.

Indefinite Franchise Term

A somewhat different problem is presented when a franchise does not specify any definite term. Such a franchise naturally will not supply any guidance as to how the refranchising process is to be conducted, and leaves unanswered the question of *when* refranchising appropriately may be undertaken. Franchises granted while the FCC Franchising Standards were in effect were limited to a maximum duration of 15 years, and earlier franchises were required to comply with the Standards by no later than March 31, 1977. Accordingly, all franchises granted before that date were required to include a definite expiration date—which cannot be later than March 31, 1992.

The effect of a pre-1977 franchise's failure to specify a definite term despite the Standards is unclear. On the one hand, as discussed above, a city may not be able to treat the franchise as if it had been amended to comply with the Standards, since the Standards were not self-executing. On the other hand, the FCC's policy of limiting franchise duration may have been sufficiently strong, and the consequences of allowing the franchisee to continue operating indefinitely so harsh, that such a franchise might be held subject to the fifteen year limitation. While this result obviously would be favorable to a city, considerable patience may be called for, since the franchise would most likely not be held to expire until March 31, 1992—the fifteenth anniversary of the deadline for compliance with the Franchising

Standards. [See *Report and Order in Docket No. 21002,* 66 F.C.C. 2d 380, 406 (1977).]

Franchises granted on or after November 15, 1977, may be of unlimited duration—at least insofar as FCC rules are concerned. The city with such a franchise must look to state law in order to determine the validity and effect of an indefinite term. It is possible that under state law an agreement for an indefinite term would be regarded as terminable by either party on relatively short notice. Because of the large initial capital investment required to construct a cable system, however, permitting termination at the will of a city during the early years of the franchise would be economically unrealistic and not in keeping with the original expectations of the parties. But in the context of an attempt by a city to terminate a franchise some ten or fifteen years after construction of the system and after the system has been depreciated fully, a court might hold the franchise terminable by the city at will.

The Problem of Unfranchised Operation

Where an operator is currently providing cable service to a city without ever having been awarded a franchise, the situation is quite different. Since there is no franchise, a city's rights and options will not be limited by contractual committments undertaken long ago. Moreover, the city would probably have the right to require the operator to cease its use of cable strung over or under public property, streets, and rights of way. Pole attachment agreements with a local utility normally would not suffice to permit an unfranchised cable operator to offer service, since such activity would exceed the scope of the franchise under which the utility operates. The city would also enjoy a freer hand in conducting a franchising process since, strictly speaking, there is no incumbent franchisee and no consideration of franchise renewal. Nevertheless, the existing operator will almost certainly regard itself—and be regarded by other cable operators—as being in the same position as an incumbent franchisee, so that the city may well be faced with many of the problems that normally arise in true refranchising situations, including the unwillingness of multiple systems operators (MSOs) to bid against incumbents.

This situation must be distinguished from another increasingly common method by which broadband video services are being provided to subscribers without authorization from a city in the form of a franchise. This method is known as SMATV—satellite master antenna television. An SMATV system typically is confined to a single high-rise building or residential complex which can be wired to one or more satellite earth station receivers and to microwave receivers without having to utilize public property, streets, or rights of way. It can provide the same type of multichannel service that is typically offered by cable systems, often using a building's previously existing master antenna system to carry the signals into individual apartments. Some SMATV operators have experienced difficulty in obtaining all of the programming services that are regularly offered by cable systems. An Arizona SMATV operator, joined by the State of Arizona, has sued several pay program suppliers and cable systems, alleging a conspiracy to put the SMATV operator out of business by denying it of access to programming. [*Multichannel News,* June 14, 1982, at 1; *Id.,* July 12, 1982, at 13; *Id.,* July 19, 1982, at 17.]

Because an SMATV system does not make use of public ways, the legal basis upon which cities require that cable systems be franchised is lacking. Moreover, the FCC has exclusive jurisdiction over uses of the radio spectrum, the means by which signals are brought to the building or complex served by the SMATV system. Accordingly, attempts by local authorities to require that SMATV systems be franchised have generally been struck down. [*New York State Commission on Cable Television v. FCC,* 669 F. 2d 58 (2nd Cir. 1982).] But where an operator makes even minimal use of public ways, such as stringing a wire over or under a street to connect two apartment buildings, a city may require a franchise despite the operator's self-characterization as an SMATV system. [*Omega Satellite Products Co. City of Indianapolis,* No. IP 82-339-C (Denial of preliminary injunction), *appeal filed,* (S.D. Ind. March 26, 1982) N.O. 82-1539 (7th Cir. 1982).] The decisions have not been uniform. An apparently contrary decision in *Meridian Charter Township v. Roberts,* 114 Mich. App. 803 (1982), which held that a master antenna system which offers a full range of programming falls within the definition of a cable television system and thus is a public utility under Michigan law. As such, the system could be prevented from operating without a franchise, even if no use is made of public streets or ways.

The presence of SMATV systems can have a significant effect upon a city's attempt to issue a new or renewal cable franchise. Where a city is as yet unfranchised, SMATV operators may acquire a head start that will be difficult for a cable operator to overcome. This could diminish the quality of the proposals that prospective franchisees will offer the city in response to a request for proposals ("RFP"), and could even dissuade them from responding at all. For example, in the current (1982) franchising proceedings in New York City, the plan of Co-op City, a huge multi-high-rise complex, to establish and SMATV system has cast doubts upon the City's ability to secure a franchise for the Bronx (where Co-op City is located) on the terms that had been offered before those SMATV plans became known.

[See *Multichannel News,* May 10, 1982, at 10]. The result may be the sort of "cream skimming" that federal policy and most cities have long sought to avoid. New York City, for example, has long pursued a strong policy against "cream skimming." [See, e.g., *Cable Television Report and Order,* 36 F.C.C. 2d 143, 208 n. 76 (1972); Mayor's Advisory Task Force on CATV and Telecommunications, *Report on Cable Television and Cable Communications in New York City* (1968).] Pursuant to this policy, the City has insisted that all unwired areas be franchised simultaneously, and that each operator be obligated to extend services throughout the area of its franchise.

Where a city is already served by a low-capacity cable system that is in need of substantial upgrading or rebuilding, there is also a danger that SMATV systems will spring up to offer higher-capacity service before the cable system can be upgraded. This may affect the cable operator's willingness to make the investment necessary to increase the system's capacity, and may deter other companies from competing for a new franchise if the city issues and RFP. Thus, delay in commencing and concluding the refranchising process may diminish or preclude a city's ability to achieve its objectives. Indeed, a city may wish to consider extending or renewing a franchise well before its scheduled expiration in order to secure immediate upgrading of the existing system.

Another threat to the franchising or refranchising process posed by SMATV is that owners of multiple dwellings may demand substantial payments for allowing a cable operator to wire or rewire their buildings. With SMATV on the scene as an alternative source of multichannel video service, this is a credible threat—at least for larger buildings. Indeed, some landlords may wish to construct their own SMATV systems and exclude cable operators totally. Either a "bidding war" between cable operators and SMATV operators or construction of landlord-owned SMATV systems could be highly detrimental to a city's ability to obtain a high-quality cable system.

Some states and cities have dealt with this problem be enacting statutes and ordinances requiring owners to allow cable operators to wire their buildings. Such laws have the effect of inhibiting the spread of SMATV, since a landlord or SMATV operator may be unwilling to invest in constructing an SMATV system, knowing that a cable operator would later have assured access to the building. New York's law [N.Y. Executive Law, 828 McKinney Supp. 1982] provides that the building owner is entitled to "reasonable" compensation as determined by the State Commission on Cable Television, which set a nominal rate of one dollar per building. The statute was challenged by the owner of an apartment building and the Supreme Court held, in *Loretto v. Teleprompter Manhattan CATV Corp.* [—U.S.—, 102 S. Ct. 3164 (1982)] that a cable operator's use of private property, under state authorization, to attach cable and related equipment constitutes a "taking" under the fifth and fourteenth amendments for which just compensation is required.

Loretto does not mean that all such mandatory access laws are unconstitutional. Indeed, the Court did not strike down the New York statute itself, but remanded for consideration by the state courts of "(T)he issue of the amount of compensation that is due." [*Id.,* at—, 102 S. Ct. at 3179.] What type of compensation scheme will pass constitutional muster still remains to be determined. Unless the level of access charges is quite low, whether determined by a flat rate, a flexible formula, or individual building-by-building assessments, the impact on the cost of providing cable service could affect the willingness of cable operators to wire or rewire less affluent areas of a city.

Review of State and Local Law

The significance of state and local law in the refranchising process may be substantial. All relevant provisions of the state constitution, state statutes and administrative regulations, if any, as well as the local charter and applicable municipal ordinances should be reviewed. In some states there are constitutional or statutory limitations on the powers of municipalities to award franchises. For example, a state may require that all franchises be non-exclusive or that they not exceed a specific maximum term. Eleven states have established regulatory commissions which regulate the local franchising and refranchising process. Some regulate cable through their public service or public utilities commissions (e.g., Connecticut, New Jersey, and Rhode Island), while others set minimum standards for local franchising. Local charters and ordinances may affect the refranchising process in a similar fashion; requirements for public notice, public hearings, or even for referendum approval must be complied with.

EVALUATING AND PURSUING A CITY'S OPTIONS

As discussed earlier in this book, a city embarking on the refranchising process has a number of options as to how to proceed: negotiating a renewal franchise with the incumbent operator; considering applications from one or more other companies where the incumbent operator may or may not be allowed to bid against other applicants; and considering converting to municipal or cooperative ownership of the cable system.

Renewing the Franchise

Franchise renewal has been by far the most common means of refranchising. A negotiated renewal offers the city a number of potential advantages. The renewal process not only is simpler, faster, and cheaper than comparative bidding, but also is less fraught with legal perils for a city. When a city entertains applications from a number of companies, its risk of being sued naturally increases. Unsuccessful bidders have attacked franchise awards on such grounds as procedural unfairness and antitrust violations. Thus, an unsuccessful bidder for the Pittsburgh, Pennsylvania, franchise challenged the award on unfairness grounds [see *Broadcasting,* March 17, 1980, at 78], but settled the suit over a year later [see *Multichannel News,* August 3, 1981, at 11]. In Affiliated Capital Corporation, the City of Houston [F. Supp. 991 (S.D., Tex. 1981)], a losing bidder brought an antitrust suit and won a multimillion dollar jury verdict. The trial court, however, entered judgment for the defendents not withstanding the verdict, and the plaintiff has appealed to the Fifth Circuit Court.

A negotiated franchise renewal avoids the legal difficulties of selecting from among rival bidders, but might still be subject to antitrust attack by companies who claim they were foreclosed from competing for the franchise. Such an attack could be made under the rubric of a conspiracy in restraint of trade or refusal to deal [Cf. *Radiant Burners, Inc., v. Peoples Gas Light & Coke Co.,* 364 U.S. 656 (1961)]. This point has yet to be litigated, but it would appear that a city is entitled to deal solely with the incumbent. By analogy to long-term supply contracts in the private sector, this approach may be viewed as reflecting the ecomomic realities of the marketplace and the reasonable expectations and desires of the parties for a degree of stability in their relationship [*Tampa Electric Co. v. Nashville Coal Co.,* 365 U.S. 320 (1961)].

Moreover, renewal of a franchise without entertaining competing bids was recognized as appropriate by the FCC when it adopted its Franchising Standards. The Commission's position survives in the form of "recommended" procedures [*Report and Order in Docket No. 21002,* 66 F.C.C. 2d 380, 387-90 (1977)]. Although the FCC did not address potential antitrust issues expressly, it has long been established that the antitrust laws must be applied in light of the Commission's actions pursuant to the comprehensive regulatory scheme of the Communications Act [see *FCC v. FCA Communications, Inc.,* 346 U.S. 86 (1953)]. The RCA Decision is not directly applicable to cable, since it dealt with common carrier regulation. In any event, FCC regulation could not confirm complete antitrust immunity. [See 47 USC § § 313 (a); 314 (1976).] It seems unlikely that a city which complies with the FCC's "recommended" franchising standards and procedures would be found thereby to have violated the antitrust laws.

Where a franchise gives an operator a right to renew, it seems still clearer that a city would not violate the antitrust laws by attempting to renegotiate terms with the incumbent. The city presumably would be obligated to deal with the incumbent to the exclusion of other potential franchisees, at least in the first instance.

However attractive the terms of a negotiated renewal may appear, a city can never be certain that a comparative proceeding would not have yielded a superior franchise. This uncertainty may expose local officials who award the renewal to criticism from political opponents and from the public at large. Indeed, in almost any community a segment of the population will be dissatisfied with past cable service and hence likely to be critical of a franchise renewal. For example, citizens in Raleigh, North Carolina, opposed the City Council's decision to negotiate a renewal franchise with the incumbent operation and mounted a petition drive seeking a city-wide referendum on whether competing bids should be obtained [*Raleigh News and Observer,* Oct. 31, 1982, at 1A]. Particularly in a city where such dissatisfaction is widespread, local officials may feel safer seeking competing bids.

Although cities have begun to show a willingness to consider other alternatives seriously, most franchises expiring in the near future are likely to be renewed. Unless a city has been dissatisfied with past service or an incumbent is unwilling to agree to significant upgrading of its system, renewal will usually be the easiest and best course for a city to follow. This pattern is likely to change only when MSOs become willing to compete for franchises against incumbents seeking renewal.

Dealing with Third Parties

If a city selects to negotiate a renewal franchise with an incumbent operator, it may never face the problems of dealing with third parties. It may, however, be approached by one or more companies that desire a franchise, or it may wish to solicit such applications.

Third-Party Initiatives

A city that elects to negotiate a renewal with an incumbent franchisee may nevertheless find itself confronted with unsolicited expressions of interest from one or more other companies. Since MSOs generally adhere to a policy of not competing against incumbent MSOs in refranchising situations, such expressions usually will be from firms newly formed by local citizens. For example, such a local group recently sought to compete against the incumbent in Santa Barbara, California. [See *Multichannel News,* June 29, 1981, at 3; August 3, 1981, at 8; September 28, 1981, at 5.] A city wishing to entertain proposals of this kind may issue an RFP or even negotiate directly with such applicants, although the latter course may present antitrust problems.

Such expressions of interest from third parties are by no means always welcome. A city that is satisfied with the performance of its incumbent operator and with the renewal terms it is proposing may prefer to rebuff initiatives from other would-be franchisees. In Santa Barbara, California, the City awarded a renewal without formally entertaining a competing proposal. [See *Multichannel News,* September 28, 1981, at 5.] In Marquette, Michigan, on the other hand, the City, following issuance of an RFP, awarded the franchise to a new locally owned firm. [See *Multichannel News,* October 18, 1982, at 1.] The author is currently serving as a consultant to the City of Marquette, Michigan. As discussed above, a city probably would not incur antitrust liability by dealing solely with an incumbent, but it may risk antitrust or other liability if it does not treat all third parties equally. A city thus should either reject all expressions of interest from third parties and deal exclusively with its incumbent franchisee or initiate a competitive procedure in which all interested parties may participate on an equal footing.

A city should also bear in mind that the renegotiation process may not run smoothly, and that changes in course may become necessary. It thus would be unwise for a city to "burn its bridges" while the process is ongoing. Careful records should be kept of all expressions of interest so that all interested parties can be notified in the event that a city later decides to issue an RFP.

Issuance of an RFP

The usual procedure in initial franchising is the issuance by a city of an RFP, followed by evaluation of the bids received and selection of the applicant to be awarded a franchise. An RFP initiates a "public proceeding" in which citizens can participate and affords fair consideration to all those who seek a franchise. In refranchising as well, a city wishing to consider companies other than its incumbent operator usually will do so by issuing an RFP. The foregoing benefits also are available in a refranchising proceeding, but the process will inevitably be complicated by the presence of an existing cable system and an incumbent operator.

In refranchising, an RFP could be issued at any of a number of junctures. Indeed, it would be possible for a city to issue an RFP at the very outset of its refranchising activities, even before weighing the desirability of renewal. This might induce an incumbent to offer better renewal terms than it would without the pressure of competition, and competitive bids could provide a standard against which a city could measure an incumbent's offer. Nevertheless, in most situations it would be unwise to issue an RFP at such an early stage. Because of the reluctance of MSOs to submit bids in competition with an incumbent MSO, a city would run the risk of receiving no bids worthy of serious consideration and thus having to negotiate a renewal without the bargaining leverage usually provided by the possibility of issuing an RFP. Finally, since issuance of an RFP before considering renewal of the existing franchise might violate a city's contractual obligations, the city should determine the extent of those franchise obligations before such action is taken.

A somewhat more likely scenario is for an RFP to be issued after a city has obtained an unsatisfactory renewal offer or has been unsuccessful in its attempts to negotiate a renewal. At this stage, a city could shape an RFP in light of its evaluation of the incumbent's past performance and renewal proposals, as well as the city's determination of its own priorities for future cable service. In this situation a city usually will defer making a final decision on renewal, so that the incumbent can submit an offer in competition with others who respond to the RFP. This gives a city the widest possible range of choices, and may also have the effect of generating a better offer from the incumbent than it had been willing to make in direct dealings with the city. Issuance of an RFP in this manner also may insulate city officials from political and public criticism that might accompany a decision to award a renewal without considering all alternative possibilities.

It would also be possible for a city to issue an RFP after it has reached a decision not to renew an incumbent's franchise, thus barring the incumbent from responding to the RFP. This procedure might induce other MSOs to submit proposals, since they would not be competing against an incumbent, but it is not at all clear how they would perceive the situation. Indeed, an MSO might be reluctant to deal with a city that has demonstrated its willingness to oust an incumbent operator, fearing that the same fate might befall it in the future.

The exclusion of an incumbent from a competitive proceeding may embroil a city in litigation. An ousted franchisee might attack a city's action as violating the existing franchise, as depriving the incumbent of fair treatment, or even as violating the antitrust laws. If a city has carefully observed all requirements in a franchise relating to review and renewal and has given fair consideration to an incumbent's past performance, it should be able to defend such litigation successfully. The antitrust laws do not require one who is dissatisfied with the performance of the other party to a contract to extend his relationship beyond the contract's expiration.

Where a city's dissatisfaction is not with the incumbent's past performance, but rather with its renewal proposal, exclusion of the incumbent from bidding in response to an RFP presents a more serious problem. Even if the bid ultimately selected by a city is superior to the previously rejected renewal offer, the incumbent can argue that it was not afforded the same opportunity as other firms and that it would have improved upon its prior offer if given that opportunity. To minimize its exposure to the risk of such liability, a city thus should not exclude an incumbent from bidding in response to an RFP unless it has found the incumbent to be unqualified. Moreover, that finding should be made in a proceeding which affords the incumbent an opportunity to be heard and all other rights of due process.

Once a city has issued an RFP, it should be meticulously even-handed in its treatment of actual and potential bidders. The RFP should specify procedures for resolution of bidders' requests for clarification of the RFP's requirements. A formal procedure of this type will help avoid *ex parte* contacts with city officials or consultants that could create the appearance of procedural unfairness. In fact, many cities adopt the salutary practice of expressly forbidding all *ex parte* contacts by actual or prospective bidders, except in accordance with procedures specified in an RFP.

A city that has issued an RFP in a refranchising proceeding faces the special problem of how to deal with its incumbent franchisee without undermining the fairness of the proceeding and thus exposing itself to potential antitrust or other liability. The city's ongoing relationship with the incumbent under the existing franchise makes some contact between the two inevitable, but care should be taken to avoid discussing matters relating to renewal of the franchise. The antitrust laws probably require a city to give equal consideration to all bids.

One issue relating to the refranchising proceeding that may have to be discussed with an incumbent is that of continuity of service. Unless the franchise obligates the operator to continue service following expiration of the franchise term—a highly desirable provision which a city should endeavor to include in any franchise—it may be necessary to negotiate with the incumbent on this matter, but care should be taken to minimize the impact of such negotiations on the decisionmaking process.

The concern of cities about potential antitrust liability has increased as a result of the Supreme Court decision in *Community Communications Co. v. City of Boulder* [—U.S.—, 102 S. Ct. 835 (1982)] holding that cities are not automatically immune from the antitrust laws in their regulation of cable television. Previously, most observers had assumed that the "state action" antitrust doctrine immunized city governments and officials for actions taken to promote the public welfare. [See *Parker v. Brown,* 317 U.S. 341 (1943).] The *Boulder* decision did not go so far as to eliminate the "state action" doctrine entirely, however, and it did not address at all the extent to which traditional franchising or refranchising activity might subject a city to antitrust liability. The rather alarmist reaction to *Boulder* by local officials [see *Cablevision,* March 22, 1982, at 63], thus may exaggerate its importance.

The narrow issue decided in *Boulder* was whether a general state "home rule" constitutional provision sufficed to immunize city action, a question answered in the negative because the provision did not represent a "clearly articulated and affirmatively expressed" state policy to displace competition in favor of regulation. The Court made clear that where such a state policy exists, local actions taken pursuant thereto may qualify for immunity. Thus, in a state such as New York, which has adopted a complex regulatory scheme for supervision by a state agency of local franchising activities, "state action" immunity may be available and unaffected by *Boulder.* Not surprisingly, many states are considering enacting legislation to overcome the effect of *Boulder,* but it is less than clear how this best may be accomplished.

The most troublesome aspect of the antitrust theories being advanced in the recently commenced litigation is the notion that a city may be obligated by the antitrust laws to award a franchise to any applicant that satisfies minimum financial, technological, and other requirements. This would treat city streets and public ways as "essential facilities" which may not be withheld from those who require access to them for entry into the cable television market [see *United States v. Terminal Railroad Association of St. Louis,* 224 U.S 383 (1912); *Associated Press v. United States,* 326 U.S. 1 (1945)]. It is by no means clear that this theory will find support in the courts. Competition among cable operators in the same geographic area is economically infeasible owing to cable's capital-intensive nature, and could lead to destructive price competition and the collapse of all but one company which then would enjoy *de facto*—and unregulated—monopoly status. [See A. Pearce, R. Peterson and M. Fredrickson, *Competitive Cable Franchising: Analysis of Economic Theory and Empirical Data* (1982).] The "natural monopoly" of cable television thus may make local franchising of a single operator pro-competitive rather than anti-competitive in its effect. Conversely, since the only real competition in the cable industry today is that among competitors for franchises, undermining the franchising process could deprive cities of the benefits of competition.

As noted, *Boulder* did not address the issue of a city's substantive antitrust liability for franchising actions. Such issues will be addressed in future proceedings in the lower courts in the *Boulder* litigation and are likely to arise in future litigation. There is little reason for cities, on the basis of *Boulder,* to abandon their well-established practice of awarding cable franchises. There is, however, good reason to exercise a greater degree of caution in the manner in which they conduct franchising and refranchising proceedings.

ALTERNATIVES TO PRIVATE OWNERSHIP

A final refranchising option is to convert from private ownership of a city's cable system to municipal or cooperative ownership. Only a small fraction of cable systems are currently operated under these forms of ownership, but an increasing number of cities are giving them serious consideration. The advantages and disadvantages of these ownership forms are discussed in detail elsewhere in this volume, but a few observations are appropriate here.

Conversion to municipal ownership can be brought about either by acquiring an existing physical plant or by constructing an entirely new system. A city considering such a conversion thus must determine whether it makes economic and technological sense to upgrade the existing system. If it wishes to pursue that course, the expiring franchise should be reviewed to determine whether it gives the city the right to purchase the assets of the present system, and on what terms. Although the FCC has never mandated such provisions, it has long urged their inclusion in local franchises and many franchises contain such a buy-out clause. The desirability of exercising a purchase right may depend upon the price formula mixed in the franchise; a "market value" provision may produce a very different purchase price than one which uses "book value" or some other formula.

Even if an expiring franchise does not grant a city the right to purchase cable system assets, it may well be possible to effect a negotiated purchase from the incumbent operator. The operator is likely to resist agreeing to such an arrangement, however, since it will usually prefer to continue its operations. It thus may be necessary to convince such a reluctant operator that the city is firm in its intention not to renew the franchise and to take over operation of the system.

Finally, if a city is neither entitled to purchase the system nor able to bring about a negotiated sale, it may be able to acquire the system's plant by the exercise of eminent domain. This course of action should be approached with great caution since, unlike the foregoing means of acquisition, the "purchase" price will not be known in advance. The applicable local law should be researched carefully to ascertain whether it permits such an exercise of eminent domain and to determine the likely acquisition cost.

If a city determines that continued operation of the existing system would be infeasible, it would be necessary to plan for the construction of a new system to replace it. In addition, the city would have to deal with the problem of avoiding interruption of cable service.

A potential obstacle to municipal ownership is the threat of antitrust liability to companies (including the incumbent operator) who seek a franchise but are turned down and thereby precluded from competing with the municipal system's monopoly. Since cable television has not historically been provided as a municipal service, a private company whose application for a new or renewal franchise has thus been denied may contend that the city's municipal cable system does not come within the ambit of the "state action" antitrust doctrine.

Many other significant legal issues will arise in connection with financing the acquisition or construction of the system and providing for its management and operation. A particularly troublesome problem is that of insulating a city from potential First Amendment problems created by its direct control over programming content. Such problems may be ever more severe than for municipal broadcasting stations, because of cable's *de facto* monopoly status. A municipally owned cable system thus makes a city government the sole "gatekeeper," determining which services and programs to make available.

Awarding a franchise to a consumer cooperative would offer a city some of the advantages of municipal ownership—for example, lower subscriber rates—while diminishing some of the constitutional and legal problems that arise when a city operates a cable system. In addition, such a system could be constructed without a city having to issue bonds to raise capital. It is doubtful that sufficient government funding for such cooperatives will be available to make this a realistic alternative for many cities in the near future, however.

CONCLUSION

At the very outset, a city must determine its rights and obligations under an existing franchise and assess the impact of state and local laws and regulations. These matters may have a profound effect on what refranchising options are available to a city.

A negotiated renewal poses fewer legal problems than other alternatives, but may not be entirely free from antitrust attack. Dealing with third parties involves greater legal perils and thus requires great caution. To ensure fairness, a city must treat all bidders even-handedly and be circumspect in dealing with its incumbent franchisee.

The conversion of a cable system to municipal ownership also raises legal and other problems. Conversion may be accomplished through purchase or condemnation of the existing system or construction of a new one, with each option presenting its own set of legal questions. There are also significant antitrust and First Amendment problems to be addressed.

There are, of course, a host of other legal issues affecting cable franchising, refranchising, and regulation. Moreover, because cable television is a growing industry, the number and complexity of such issues will likely continue to increase in the years ahead.

Rex Reynolds

4. Franchise Renewal: A Municipal Perspective

The ink was hardly dry on the birth announcement of cable in the United States when Dubuque, Iowa, got its cable system in 1954. Because of its unglaciated, limestone hills and its considerable distance from the nearest television station, consistent and dependable television reception was minimal, even from a translator erected on a high hill just across the river from the town. So cable was a natural for Dubuque. And like most systems and franchises of the early 1950s, Dubuque's cable system and franchise were primitive by today's standards. Responding to the cable system's request for a rate increase in the early 1970s, Dubuque tried its luck at the cable slot machine and hit what it thought was a jackpot. With the able assistance of cable consultants, an amended cable ordinance granted a rate increase to the cable company. The new ordinance required an access studio, a cable commission for the citizenry, and a franchise fee for the city.

But the studio space at the local cable office, though reasonably suitable, was never considered adequate by local producers. The studio equipment was, for the most part, used or surplus pieces from cable company storage closets; maintenance was sporadic to poor; repair was seldom; and downtime was the rule. The cable commission was empowered to meet, to listen, to give advice, and to administer the fee monies received from the cable company. Otherwise, the commission lacked authority to correct problems. Finally, the fee for the city was five cents per month per

---

Rex Reynolds is Professor of Speech Communications, Loras College, and former Chairman of the Dubuque, Iowa, Cable Commission.

subscriber—a total of two-tenths of one percent!

Dubuque had tried to make strides for local residents in terms of service and fees—and lost.

The only reasonable hope for an effective cable system in Dubuque was through the refranchising process due in 1982. The cable commission became determined not only to hit the jackpot at refranchising time, but to win. To do that the commission knew that it could rely on neither chance nor luck. The four essentials to true winning at refranchising are: (1) super preparation, (2) political competence, (3) unassailable franchise language, and (4) effectual negotiating. When it came, the refranchise pay-off was a true win, worth every moment of hard work!

INGREDIENTS FOR SUCCESS

Cable Education

Early on, the commission members thought that they would eventually engage consultants to carry the ball for them. After nearly three years of intensive preparation, the commissioners realized that their own cadre of "cable experts" would be more than adequate for the job. The franchise fee which the commission had managed to get in the amended ordinance of the early 1970s averaged $8,000 to $10,000 per year; while much of it was allocated to promote public access, a portion was used to provide cable education for the refranchising process. The commissioners subscribed to and digested virtually every trade journal related to cable. They purchased and read known cable titles; collected cable franchises and Requests for Proposals (RFPs); attended several National Cable Television Association (NCTA) national conventions with carefully planned agendas of their own; joined subscriber, access, and municipal organizations; and took part in several seminars and workshops. While these activities generated volumes of notes, looked like boondoggling, and raised a few eyebrows at City Hall, the refranchise results more than justified the monetary and voluntary-time investments.

Supplementing such directed cable education, the commission logged subscriber complaints and operator reactions. Later evaluation and classification of these lists, covering several years of experience, produced an encyclopedia of what could possibly go askew in a cable system, a litany of cable company reactions and strategies, a fantastic base of pertinent franchise issues, and a very useful profile of local cable concerns. The commissioners listened, read, and prepared well.

Political Competence

The second essential in Dubuque's refranchising experience is political competence. The two sides of the local political scene are politics within the community, including the operator's posturing and lobbying, and the cable corporation itself.

In the community, the ultimate approval for the franchise required approval by the city council and a referendum of voters, as required by Iowa state law. Gaining support from city staff and community leaders and spokespersons, and soothing the concerns of cable aficionados and subscribers were all part of the preliminary process. In Dubuque the cable commissioners were generally subordinate to the city staff. Subscribers and those with an interest in cable technology or access flooded the commission meetings with their complaints and advice. Tiptoeing amidst the real and imaginary turf is a ballet that cannot be taught or described in this piece.

Neither can the trade secrets be easily revealed on how to penetrate the cable company's corporate veil enough to predict and prepare for its moves, responses, and strategies. Suffice it to say that the approval from a multiple systems operator (MSO) itself would depend on the internal structure of the cable corporation, its personnel, and how all these players perceived themselves, interacted, and sometimes made their presence felt in the community. Likened to slot machines, often the hollow reverberations of franchise tokens cascading into the metal basket, and the flashing lights and ringing alarm accompanying a cable jackpot, are orchestrated by cable interests to induce a local thrill and mask a loss by the municipality.

In the end, the recommendations of the cable cadre were strongly approved by the staff, the city council, the voters, and the corporate officers, in spite of frequent objections and tense moments. Political competence proved to be a most significant essential!

Franchise Language

In legalese, writing a document is properly referred to as drafting. "To draft" means to pull out, select, draw, design. Indeed, the usual method of drafting a will, contract, or articles of incorporation is to begin by selecting passages from existing documents, better known as "boilerplate." This selection is justified when the language is legally unassailable or virtually unassailable because the passages have a definite meaning without variation, usually as a result of legal requirements and court interpretations. Since it is an easy, quick, and inexpensive process, passages from existing cable franchises often are presumed to be boilerplate with definite meaning, and are selected for new franchises. However, there is little basis for supposing that cable franchise language is boilerplate, tried and true, unassailable. As of this writing, no comprehensive study of the legal meanings and interpretations and applications of cable franchise language has been published. Among the reasons why cable franchise language has not met the boilerplate test are the following: the relative newness of cable franchises, the constant flux of federal cable regulations, the diversity of state laws regulating or ignoring cable, the absence of published lower court decisions related to cable franchises, the uncertainty of current and future directions in Congress and the FCC, and the lack, until recently, of a unified direction among municipalities.

The Dubuque cable commission not only collected franchises, but also reviewed the cacophony of FCC decisions, inquired into lower court cable cases, interpreted and applied federal court decisions, investigated other communities' interpretations of and experiences with cable language, and sought advice from attorneys with cable experience. While many passages in the Dubuque franchise are borrowed, all passages, clauses, and sections were subjected to the rigorous scrutiny of research, review, and redrafting—a process few communities undertake—a process donated to Dubuque as a labor of love.

A major strategy adopted by the Dubuque cable cadre was to negotiate only from language they had processed or that was in direct reply to such language. The MSO submitted its first proposal before the commission had completed its RFP; that proposal was received, filed, and never discussed. A second proposal, responding to the approved RFP, was rejected and declared non-negotiable, partly because the cable cadre wanted to negotiate only from its carefully-tested language. While time and use will ultimately test the unassailability or vulnerability of the franchise language, Dubuque was not leaving its win to chance.

Effectual Negotiating

The fourth essential in the Dubuque refranchising experience was effectual negotiating. Of course, negotiating is a fact of daily life, and everyone is an experienced and successful negotiator. The essentiality of effectual cable negotiating lies in preparation, a grasp of communication laws and regulations, understanding telecommunication technologies and trends, knowledge of community needs, the political complexities and competencies, opposition positions and strategies, skill and experience of the team, and the corresponding strengths and weaknesses for the opposition. Since very few effectual cable negotiators are available to municipalities, the Dubuque commission and council elected to go with their local cadre of "cable experts" who then negotiated the franchise.

* * * * * * *

When the refranchise countdown was over, two weary negotiators came home to Dubuque with a new franchise, designed for and by Dubuquers, which was indeed a winner. The terms of the franchise speak for themselves:

While Iowa state law requires a franchise to be enacted by a city council, frequent references in the Dubuque ordinance emphatically state that the document is "a mutually bargained and negotiated contract acceptable to both parties." No unilateral heavy handedness; no antitrust; no litigation about rights or privileges given away.

Cable services are to be available to all everywhere in the city, wherever its boundaries. If service is interrupted, cost to the subscriber will be adjusted. Service and repair are backed up by a 24-hour hot line.

Basic service, at $7.25 per month, include 60 activated channels (30 now, 47 in five years, all 60 in ten years), a band full of FM radio, two outlets, and a converter. Discounts are available for annual payments, elderly subscribers, and buy-in contracts. Additional channels for Pay TV bring the system up to a minimum of 70 channels.

The city gained the right to require types of programming, specific two-way services, improved standards, and control of the government, health, education, religious, and community programming access channels. A separate closed-circuit institutional loop will serve the schools, hospitals, and city and county buildings.

Specific amounts of money are required to be spent for construction (two offices), operating expenses, two-way services, access assistance ($439,000 first five years), access studios (two video, one radio access), mobile units, access equipment ($660,000 initially), local origination ($2,555,000 first five years), additional satellite dish antennas, back-up equipment, converters, and maintenance.

The city receives a five percent fee on all gross revenues, paid in advance each year, plus control of $2,000,000 in special update funds for adding advanced technologies and services not comtemplated in the franchise.

Privacy invasion, electronic surveillance, and unwanted release of personal data are controlled. Open books and detailed reports are required.

The city has the right to interpret or waive any provision, and has maximum flexibility to adjust to future deregulation or increased regulation. For full compliance the city can resort to a range of procedures and penalties. Any litigation must go through local, state, and federal courts with costs and attorney fees the burden of the loser.

And what does the franchise grant the cable operator? In a city of 22,000 homes with a cable penetration rate in excess of 80 percent, the coveted right to sell an essential service to the public protected, for all practical purposes, from cable competitors, and the privilege of using public property for its plant, the MSO has the endorsement of the city and of key leaders for its product, the assistance of access programs integrating it into the community as a status medium, and documented proof of its good business judgment and ability to operate one of the most outstanding systems in the nation.

The local "cable experts" firmly believe that what's best for the community is best for the franchise holder. So, the MSO can declare a win, too!

Rod Thole

5. Franchise Renewal: An Operator's Perspective

Other chapters in this book discuss what cities might ask of the cable television operator at renewal time. The pages that follow will approach renewal by concentrating on some of the issues that confront a cable television operator. By doing so, perhaps some appreciation can be developed for the anxieties faced by the operator at this very critical time in his business relationship with the city. By understanding the operator's perspective, franchise authorities may gain insight on how to approach and administer the renewal process, thereby making the experience less traumatic for themselves and the cable television company.

As the last of the original franchise-granting dwindles, system franchise renewal looms as an issue of major proportions for the cable television industry. The process tests anew the often fragile relationships that exist between cable company and the community. It will also test the business ethics and professionalism practiced among competing companies. Renewal can be a frustrating and nonproductive experience for the operator or it can be extremely positive, resulting in improved cable TV services and a renewed commitment to the city and continuance of quality programming and technical services to subscribers. If both operator and city approach the process with mutual respect, cooperation, trustworthiness, and a genuine desire to reaffirm the partnership in the best interests of the city, citizens, and cable TV company, refranchising will properly commit all parties to a

---

Rod Thole is Senior Vice President, Telecommunications Group, Heritage Communications, Inc.

renewed enthusiasm for the tremendous benefits to be gained from a sound cable TV operation. If unrealistic expectations, mistrust, and a lack of communication prevail, the process will quickly become antagonistic and burdensome to both parties. Expensive lawsuits, polarization of community groups, operators held hostage by a city's renewal demands, and a general disruption of cable television service will all be part of an unpleasant and prolonged renewal battle. No one gains in this situation and both the company and the city must come to renewal negotiations agreed that a controversial stance must be avoided.

Experience demonstrates that the cable television business is risky and capital intensive, money currently is tight, and operators are facing stiffer operating requirements. Some operators will begin selling off older systems because of the expense in carrying extended loans with reduced cash flows. Urban cable development will be slower than expected, since operating margins have diminished, capital costs of proposed new systems are excessive, and franchise renewals are likely to require expensive upgrading with relatively unattractive return-on-investment prospects. This has sent shivers through the investment community and is creating legitimate financing problems for the industry. Cities need to understand and appreciate these changes as they develop their renewal expectations.

Responsible operators are taking a new look at past proposals such as low-cost basic service, expansion of channel capacity, addressability, the programming explosion, advertising, and cable text. At least for the short term, they see expenses looming larger than revenues in the introduction of these services. Confirming this gloomy forecast are the financial analysts who have expressed alarm about shrinking operating margins associated with cable television systems that are upgrading too aggressively. Wall Street is also concerned with uneconomically low rates for basic services and the high cost of new franchises and refranchising. Cities must be willing to face the hard financial truths of today's cable television industry and be willing to accept realistic and financially-sound renewal proposals.

As the renewal mood of the franchise authority becomes clear, the cable television operator may experience considerable anxiety. His business moves into a period of uncertainty; renewal negotiations may be disruptive to cable television operations, may affect morale of employees, cause fear among his lenders, place new pressures on operating and capital funds until the franchise is renewed, and generally threaten the viability and continued success of a business that may have taken years to reach maturity. Having taken the risks to establish the business, the cable television operator is often annoyed at having to defend or justify the vagaries of a business that perhaps only he can truly understand or appreciate. Large multiple system operators (MSOs) and small independent cable TV operators do, during a typical 15-year franchise period, risk substantial sums of money, they become part of the community's business and social environment, and they develop relationships with subscribers, lenders, programmers, employees, and community groups. What has taken years to build into a mature system may, as part of the franchise renewal process, now be challenged and be subjected to political posturing, unrealistic whims, a lack of appreciation or understanding of the business, and, in some rare instances, a total disregard for the economic realities of cable television operations. If a community recognizes these natural operator fears and is sensitive to them, the procedures and policies a city establishes for the franchise renewal can immediately help in putting some of these concerns to rest. To assure an orderly franchise renewal process, procedures and activities related to the renewal should begin positively and in a spirit of cooperation—well ahead of the franchise expiration date.

RENEWAL NEGOTIATIONS

During renewal negotiations, the operator will be concerned with a smooth flow of activity at the cable system. The renewal should not interfere with operations. A threatening or uncertain posture by the city should be avoided. Standards, procedures, and directives the city initiates may have impact on the daily routine of cable employees. The public attitude the city displays can be a positive or disruptive influence.

The city must recognize that the renewal process is an emotional as well as business undertaking. Dealing with the future of the cable company is also dealing with human beings. Employees will be concerned with job security, shareholders and equity partners with the safety of their investments, lenders with the impact of the renewal on financial performance and the stability of the system, and subscribers with the continuance of quality cable television programming and technical service. To eliminate or reduce these "human fears," the city and operator need to enter the renewal arena early and place all discussions in an open, candid, and honest forum.

Procedures

As a basic element of the renewal process, the operator will want to be clear on procedures. What are the renewal steps the city wants followed? What are the rules of the game? What does the city expect? Who at City Hall is in charge? Who will actually make the decision: the city council;

community groups; a subcommittee of the council? What impact will citizens-review committees play in the process? While thinking these thoughts, the operator will typically be concerned with the expense of renewal. Will a cadre of lawyers be required, will the process be long and involved or simple and orderly, will he need to develop elaborate and costly documentation of past performance and plans for the future? In short, how will renewal impact on the financial operations of the cable television system and how long will it take to complete the refranchising process?

Consultants

Consultants often play a key role in the renewal process. An operator will be concerned with a community's use of consultants. Will the past stewardship of the operator be judged by local people who have been served by the cable television system or exclusively by an outside consultant? If a consultant is utilized, will he or she understand the local problems, the history of the system, and the restrictions that may exist, or will the consultant encourage unrealistic expectations in the minds of the city council? Who will pay for the consultant? Will the consultant be knowledgable and qualified? Will realistic recommendations be made? A poorly trained consultant or one lacking proper objectivity, can cause difficulty in completing the franchise renewal process on an orderly and timely basis. In selecting a consultant, the city should make every effort to assure objectivity, solid experience, and demonstrated knowledge of cable operations and finance.

Politics

Right or wrong, politics play a role in the renewal process. The political environment may give an operator cause for alarm since political friends and enemies of the operator may introduce the potential for positive or negative bias. Retaliation for past sins or rewards for favors may become part of the review process, making the renewal a mini-political campaign. Renewal should be judged in an objective setting. A recent *St. Paul Dispatch* editorial summarized this dilemma by saying:

> There ought to be a change in attitude on the city council toward cable TV franchising. The franchise process is city business, not campaign contribution collection time or help your buddy time. Cable subscribers deserve to have

> the best possible system at the lowest possible rates. They need to believe that their elected representatives are after the same thing.

If the franchise renewal process is open and above board, the operator will recognize he must account for his past stewardship and welcome the opportunity to share information, experiences, successes, failures, and future plans through a constructive and meaningful dialogue. City officials, in turn, must be willing to reciprocate and provide an equally positive forum to foster a healthy renewal negotiation.

Competition with Other Cable Companies

An annoying and emotionally-upsetting phenomenon of recent franchise renewals is the circumstance where an established, qualified, and proven cable operator must risk his business to the outlandish proposals of an inexperienced johnny-come-lately competitor. All operators, no matter how professional they are in operations, or how committed they might be to the community, and to their subscribers, have had failures and disappointments during the franchise term. The nature of the cable TV business is such that it is difficult to please everyone all the time. While an operator may strive for excellence, few can present a flawless or perfect record to city hall. (Few city halls could post such a record to their citizen constituencies either.)

While it may be appropriate at renewal to negotiate some changes with the operator, the temptation of cities to throw open the process to all comers should be *avoided.* Harold Horn, Executive Director of the Cable Television Information Center and a major consultant to cities on franchising and the renewal process, recommends that cities start from the premise that the community is best off to reconcile itself with existing operators even if there are problems to be worked out. Mr. Horn's advice is to be heeded; if an operator has generally met the requirements of the franchise ordinance, the company should be given every opportunity to continue and to develop appropriate plans for the new franchise term.

Generally, the company interested in competing with an incumbent operator is on the periphery of the business, not in the mainstream. Often it is this inexperienced newcomer that creates problems in refranchising. Such a company may attempt to wrestle the franchise away by proposing more service, lower rates, more give-aways, more, more, more, with no real appreciation for, or understanding of, the financial consequences associated with these proposals. On the surface, these offerings may appear attractive to cities, but competing applicants, particularly those with little or no cable television operating experience, need to be carefully screened by franchising authorities. Investigate competing applicants thoroughly—what is their background, what are their financial resources, what are their qualifications and credentials to deliver what is proposed? Cities should not be deceived into thinking that these new entrepreneurs have all the answers.

If legitimate, experienced, and qualified competition enters the renewal process, the city should avoid introducing the competition in a threatening or menacing manner. A competing applicant should not be used as a means of intimidation in negotiating with the current operator. Fairness and objectivity, with hopefully some concern and appreciation for the incumbent's performance, investment, and past record of service to the community should be the order of the day.

Early Negotiations

On occasion, operators may find it necessary to approach communities for renewal years in advance of the scheduled renewal date. While the need for an early renewal may appear to be an operator advantage, closer examination of the facts demonstrate that early renewals can be in the best interest of the city and subscriber. An operator who manages his business properly may want to upgrade the cable system and expand services as needs arise well before any pressure from the community is exerted. Unfortunately, the operator may not be able to gather the financial resources to build these needed improvements with only two or three years remaining on a franchise. No bank will loan money under these conditions. Without the option of sufficient time to develop returns on additional capital investment, the operator is at an impasse with the lender and subscribers may not be served as they should be. Regardless of the lending source, a cable company must have some formal renewal assurance from a municipality before financing can be arranged and committed for future expansion. For these reasons, cities should not be reluctant to enter into early renewal negotiations.

TERMS AND EXPECTATIONS

At renewal, an operator will be interested in amending or deleting from the existing franchise those provisions which have, over a period of time, proven to be unworkable, unrealistic, or unnecessary, or because of changes in laws, have become moot. Cities should exhibit a willingness to accept such changes. Every effort should be made to upgrade the franchise agreement and ordinance to reflect current conditions. Cities may also wish to alter or amend the franchise provisions from their perspective, and few operators would object to reasonable and properly motivated changes on the city's behalf.

As a city considers a continuing relationship with an existing system, the operator must reassess his investment, the level of his operations, and the financial and operational committments he is prepared to make for a new term. Some operators may elect to sell, opting to take what financial rewards exist and turn the cable television system over to an operator who will have the financial strength to meet the demands of the city under new franchise requirements. Most operators will elect to continue to do business and to grow with the community.

During renewal discussions, a city can reasonably expect from a conscientious and well-intentioned cable company a summary of the cable system's past operating experience. This presentation might include historical data regarding subscriber growth, revenue and expense increases, operating cash flow analysis, capital investments, market surveys identifying attitudes and needs of subscribers, rate increase history, and other information that the operator has gathered and used to make business judgments during the franchise term. Enlightened operators will recognize that this information will help the city to make sound judgments on past operations and better prepare them to evaluate the renewal proposal. Of equal importance is a detailed presentation from the operator documenting system plans for the renewal term. This would discuss plant upgrades and rebuilds where appropriate, expanded programming options if feasible, other capital improvements, and a look at the rate-making requirements under a renewed franchise term.

Part of the uneasiness experienced by the operator at renewal is related to (1) unrealistic demands, (2) over-regulation, (3) a disregard by the city of cable economics and financing, (4) unrealistic and unqualified competitors who will promise the world to grab the franchise, and (5) an apprehension that the renewal process will be used as an opportunity for the city to retaliate for genuine or imagined lack of cable television service.

Expectations

A good place to begin is to recognize that during the past several years, large public companies have been ingratiating themselves into the major markets with ambitious cable TV promises, and that the fallout has spread to older, smaller communities within earshot of these "blue sky" proposals.

Competitive bidding for the few remaining large-city franchises has encouraged more modest communities to ask for 108-channel, 400 megahertz, interactive systems, with all the bells and whistles, just like the franchise granted in the big city down the road. In the eyes of municipal authorities, the high capacity, two-way systems serving up a gigantic smorgasbord of programming options and flashy ancillary services, at artificially depressed subscriber rates, tend to label smaller cable television market operations as noncompetitive.

The *high* density, large population, urban centers may be capable of supporting grandiose cable systems. But approaches of this nature in smaller communities are unwise since the maximization of cable TV plant and investment in large cities permits higher revenue streams and sharing of the operating and capital costs of the cable system over a larger subscriber base. Furthermore, lurking in the background is considerable scepticism on how well these new urban telecommunications systems will actually perform over the long run.

Municipal officials need to be aware of over-promising by cable companies and should halt the practice of awarding franchises based upon a "more is better" approach. Cities can be exposed to a great number of promises and many communities openly encourage unrealistic bidding. Granting a franchise to the "best offer" does not always mean that a city is going to get what is in the proposal. Only time will tell if promises will translate into operating reality. For these reasons, a "more is better" approach to the renewal process is not in the best interest of the city, subscribers, or operator. The level of service which can be supported in one community may not be possible in another. One of the fears of an operator at renewal time is the thought that no matter what has been done or how well the system has been operated, the city may say "that's not good enough."

Rebuilding, Upgrading, and Overbuilding

Renewals can raise the question of rebuilding or upgrading the cable system. Communities may request 12-channel systems to be upgraded to 36 channels, 36 channels to 54, 54 to 108, and on it goes. The decision to rebuild must be based on solid reasoning. If the cable system is not operating properly, or if necessary, programming services cannot be provided because of limited channel capacity, and the economic conditions exist to support an upgrade or rebuild, an affirmative decision may be appropriate. If the rebuild request is a political demand or a city's ego trip with no market support to justify its request, a rebuild may be an invitation to bankruptcy or insolvency for the cable system. In small systems, upgrades are more practical and feasible than rebuilds. It's one thing to upgrade through use of modification kits on existing amplifiers, but making improvements by removing amplifiers and cables from poles and respacing new equipment may cost many times that of the modification route. To rebuild everything, including all cable, will cost even more. The decision to upgrade or rebuild and the method employed to achieve it must be guided by market conditions, subscriber needs, and financial feasibility.

Recognizing that most franchises are non-exclusive and having discussed the question of competing applicants, a few words on the overbuild dilemma are appropriate. An "overbuild" is a situation where a community may elect to renew the incumbent operator, but simultaneously grant a second franchise to a competing company. When this rare circumstance occurs, it is often with the misguided opinion that two operators are better than one, that competition is healthy, and that one operator will keep the other honest, competitive, and more consumer-oriented. Overbuilds are disastrous for operators and rarely beneficial to subscribers. Experience ahs shown that sharing a limited subscriber base results in poor financial performance and ultimate deterioration of program and technical service for both systems. Few communities can indulge in viable overbuild options and any tendency to move in this direction must be considered very carefully. While the aspiration may be for improved service and lower rates through competition, the real-life result may be two extremely weak cable TV systems neither of which can muster enough resources to serve subscribers well.

Financing and Overregulation

The perception that a cable TV company is a money machine is wrong, and sane understanding of the true financial characteristics and economic realities of cable must guide negotiations. Cable television is a business, and it must be reasonably profitable if it is to meet its obligations to lenders, shareholders, and the future growth and expansion of the system itself. As mentioned earlier, distinctions must be made between large metropolitan cable systems and small- to medium-sized systems. Truthful and knowledgable interpretations need to be made regarding rate-of-return analysis, capital investment, and interest costs. The high-risk aspects of cable television need to be considered by cities when evaluating profits and bottom-line numbers. The speculative nature of a cable TV investment should enjoy a better rate of return than more predictable or guaranteed investment instruments. At a time when treasury bills and short term certificates of deposit enjoy high return with no risk, a 15 to 25 percent rate of return for a more speculative venture such as a cable television system would not seem unreasonable. That a cable television system may be enjoying a good profit and attractive rate of return at renewal, is not unusual. It is quite typical since, at renewal, the system has developed to maturity. All too often cities look at company financial statements only during the year of renewal and ignore the start-up years in which large losses accumulate quickly. Just because a system generates good cash flow in year 14 or 15, is not an indication that it has been profitable throughout the franchise term. The later years of a franchise is the period in which the losses of earlier years are finally paid.

Another "hidden" expense which must be considered in evaluating the cable operator's financial projections is the cost of burdensome regulation. While significant regulatory relief has been achieved at the federal level, state and local regulation of cable television remains at a high level. In a major examination of the cost of regulation, a study by the National Cable Television Association found that 22 percent of subscriber revenues are used to cover costs of municipal and federal regulation. Released in May of 1982 and conducted by Ernst and Whinney, the study analyzed a 35-channel cable system located in a major metropolitan area. This system was required by franchise to have two-way capability and madated construction of three production studios with an additional requirement for a fully-equipped mobile unit for public access. According to the study results, 21 percent to 22 percent of the system's total subscriber revenues were funnelled to cover costs created by those franchise requirements.

With less than 30 percent of television households now subscribing to cable, there is considerable potential for cable growth, but investors are discounting earlier hopes of reasonable returns from cable investments due mostly to high interest rates and the uncertain local regulatory environment. Of particular concern is whether cable and its rapidly emerging competition wil be treated with parity on the regulatory front. Renewal is an opportunity to assure that regulatory mistakes of the past are corrected and that responsible rules are established for the new franchise term.

Earnings projections in the study noted that to achieve an 18 percent rate of return, the system would have to generate $24.83 per subscriber per month. Regulatory costs ranged from $5 to $6.08 per subscriber per month. The study also said its estimates may be conservative because an additional layer of state regulation was not factored into the statistics.

Other costly requirements cities place on cable operators, declared the report, include franchise fees, local origination facilities, public access channels, and free service to schools and governments. These financial burdens are usually recovered from the system's subscriber base even though individual subscribers often don't want or use the services.

The report also listed other costs that were not included in its initial results that could nonetheless increase cable operator expenses. They include construction of institutional networks, expansion of channel capacity, prepayment of franchise fees, unreasonably tight construction schedules, high pole attachment fees, and rate increase restrictions. The study confirms that the costs of regulation can cause subscriber rates to be higher than they need to be. Unnecessary regulatory costs hurt the system's subscriber growth potential and can cause customer migration to competitive alternatives.

At a time when cities are suffering declining revenues, operators are concerned that municipal officials may seize the opportunity to develop new revenue streams through increased franchise fees. This can do serious harm to a cable television system. Cities which force or extort from cable operators fees to support municipal services, such as repair of city property, are also doing a disservice to cable television subscribers. Municipal services should be supported by all citizen users and cable television subscribers should not be singled out for additional taxes or assessments to operate city services.

Franchise fees should be reasonable and the funds collected by the municipality should be put into activities which directly benefit or enhance cable television services. Support of community programming, purchase of electronic character generators for the display of government information on the cable system, assistance in building unprofitable line extensions to serve city and public buildings, establishing production training workshops for schools and the community at large, are all examples of what some enlightened cities have done with their franchise fee proceeds.

Productive cable TV assets are amplifiers, converters, cable, earth stations, headends, and service trucks, to name just a few. These assets produce the cable television services that subscribers purchase. They contribute to the income and the revenue stream of the cable television system. Local origination studios and vans, institutional loops, public access equipment, and unnecessary channel capacity do not contribute to the income or revenue stream. They can be viewed as nonproductive assets. Operators view with regret the tendency of franchise authorities, well-intentioned citizen groups, consumer-oriented public agencies, and, on occasion, consultants, to require the operator to build or purchase nonproductive assets. If too many nonproductive assets become part of the investment requirement, the return on the investment will be lower, or worse, the system income may not be able to meet debt obligations and operating expenses, eventually resulting in a deteriorated or financially-troubled cable television system. City renewal demands need to minimize nonproductive assets and balance their need against what the cable system can financially support.

Competition with Other Technologies

We have discussed operator concerns related to overregulation and unreasonable franchise fees. These issues take on even more importance and heightened sensitivity as the competitive threat of new technologies enters the marketplace. Cities often view cable television as a monopoly. This attitude is used as the impetus for placing regulatory constraints on the cable television system. Cable television today is not a true monopoly. Unlike a utility, cable TV is not required by all users. Accepting cable television service is a matter of choice; no one is forced to take cable TV. This is not the case with electrical, gas, water, or sanitation services. Unlike utilities, the services provided by cable television are increasingly available from other sources. Cable television competes for the consumer entertainment dollar with theatrical films, broadcast radio and television, live sporting events, cultural activities, and other forms of local entertainment.

In recent years, new technologies have had dramatic impact on cable television operations. Microwave multi-point distribution (MDS), over-the-air subscription television stations (STV), video cassettes, discs and recorders, home satellite receivers, satellite master antenna television operations (SMATV), low power television stations (LPTV), the soon-to-be-initiated direct broadcast satellite-to-home systems (DBS), and new competition from a deregulated telephone industry are all competing technologies currently providing programming identical to that distributed on the cable. This new competition has been and will continue to be felt in the telecommunications marketplace. Given these new competitors, cable operators view proposals by cities to raise franchise fees, demands to add nonproductive assets, and other costly franchise and regulatory requirements as contributing to the difficulty of competing with other unregulated competitors. Additional regulation, limited only to cable TV, places cable at a serious disadvantage. It's unfair to place regulatory and financial burdens on cable that its competitors don't have. These competitors require no franchise from the city and pay no franchise fees. They are not under the regulatory control of the municipality. Cities must recognize that new entertainment delivery systems make the risk of operating cable television systems even higher than before and some consideration must be given to the competitive technological environment in establishing reasonable franchise fees, regulation, and franchise terms.

Municipal Ownership

The concerns of the cable television operator at renewal time cannot be expressed without saying a few words about municipal ownership. Franchise ordinances do, occasionally, have provisions that permit cities to purchase cable television systems at fair market value at the end of the franchise term. Some communities have toyed with the idea of owning the system looking to cable revenues to generate additional funds for city operations. There are many reasons why a city may consider municipal ownership. Most of them have to do with cable being viewed as the "golden goose." The record to date, however, suggests that most cities, after researching the matter, will come to the conclusion that cable is still a speculative enterprise in which they must place public funds at risk. Putting public finds at risk is not good public policy.

The case for government ownership breaks down further when the aspect of government encroachment in the free enterprise system is considered. Should government establish its own profit-oriented businesses at the expense of the private sector? If we are to preserve our free enterprise system in its present form, the answer is no.

Municipal ownership becomes even more unacceptable when the business contemplated by the government is communications. How would a city go about employing newscasters, writers, directors of programming, and making program choices, and other major channel-carriage decisions in a way that would bring no criticism as to fairness, bias, or political influence? Within a political process, this is virtually impossible. The dangers inherent in municipal ownership of media should be obvious. Any community seriously considering issuing to itself a cable television franchise, should carefully examine the constitutional issues involved, the high risk financial consequences for its taxpayers, the history of community referenda on public ownership, and the inevitable potential of censorship, bias, and political abuse of government control of what is essentially a fre enterprise, First Amendment activity.

SUMMARY

Hopefully these pages have developed some appreciation for the problems a cable television operator faces at franchise renewal. The best renewal approach should be one of balancing the real needs of the city, the costs to the operator, and subscriber willingness and ability to pay for additional services. If the operator is prepared and willing to take an open, honest, and straight-forward approach to examining past performance and future plans, the city should be willing to enter the dialogue recognizing that a cable television system is a delicate balance between demands of the city, subscribers, lenders, shareholders, and government regulations. The demands and pressures placed on an operator at renewal need to be further measured against the business risks and the ability for the cable system to remain a healthy "going concern," fully capable of meeting the needs of the subscriber.

Renewal will see the city concentrating on rates, service, and additional benefits to be gained from the cable system. The cable operator will identify as key issues the following: reasonable franchise expectations, fairness, economics, and an opportunity to continue to operate the cable system in a responsible and businesslike fashion. If the cable company has lived up to its commitments to the community, it is an injustice to lose a franchise at renewal. Recognizing this and remaining sensitive to the issues pointed out earlier in this chapter, cities and operators should approach the renewal process in a manner which will assure good relations and constructive discussion which ultimately will lead to continued service to the community.

We have observed that cable operators are often their own worst enemies and that their over-zealousness to win new franchises and take away old franchises from incumbent operators may result in unrealistic promises that could go unfulfilled. There are a lot of long, cold winters between the promises of today and the reality of 1995. To avoid a serious mistake, city officials should examine a company's past record, look at responsiveness to the community, review its management philosophy and its track record, and look at the financial resources available to the company.

As previously noted, successful refranchising negotiations will depend greatly on the advice cities receive, and the common sense of city attorneys and consultants. Cities must understand that you cannot do for a 10,000 subscriber city what you can do for Dallas with over 400,000 subscribers. Older markets are not going to behave like new-build situations. It's a different environment. In mature systems, there is not the appetite for multi-pay, or the willingness to use the new non-entertainment services. That is often difficult to explain to a consultant or city council.

Most important, cities need to be prepared to accept translating their franchise demands into financial consequences for the cable system. Two groups will pay for unrealistic demands—the taxpayer or the subscriber. There is a cost to the subscriber for every service and asset. The public wants an economically attractive cable service. Frills are not a priority for the paying subscriber who wants low rates and reasonable service.

As we close this operator's perspective of the franchise renewal process, it should be emphasized that in the enthusiasm for the future and in the rush to add new services to cable systems, operators and cities must avoid getting carried away on the wave of technology. While operators must promise only what they can deliver, cities should demand only what is deliverable.

Thomas T. Taylor III & Constance Brand

6. Access: The Community Connection

Access, in cable terms, started in the United States in the 1960s and early 1970s, in many ways a different world than we face today. The wave of the 1960s and early 1970s has washed over us and receded, followed by a new wave of limits, tighter budgets, and deregulation in the 1980s.

Despite these economic pressures, access has survived and prospered. Deep historic roots give it its strength. It is the fundamental belief of the democratic approach to society that each individual has worth, has an equal right with all others to voice his or her opinion on public matters, and has an equal right to hear the expressions of all other fellow citizens.

Access to the means of communication both as "writers" and "readers" is fundamental to our First Amendment tradition. Until the advent of access to cable, exercise of such rights on television, the most powerful and pervasive means of communication in our society, was severely limited.

It is a cliché to call our time the age of communication or the information

---

Thomas T. Taylor III is a member of the Board of the Portland Cable Access Corporation and is a producer and educational consultant in the use of community media. Previously he was a Professor at the Center for the Moving Image, Portland State University.

Constance Brand, independent film and video producer, currently is a resident artist with the Video/Filmmakers in the Schools, Northwest Film Study Center, and the Northwest Regional Coordinator for the National Federation of Local Cable Programmers.

age. Problems from domestic violence to insurrection are often seen as "failures in communication."

Although telecommunications in general and access in particular are not going to overcome "failures in communication," they can help the process, and they can make a profound difference to informed decision-making. For this reason, a city undergoing the process of refranchising or re-negotiation of the franchise must take the process seriously and look at it in terms of what it is fundamentally about—not just economics, not just politics, but an important means of making our society work.

CHARACTERISTICS OF ACCESS

Access has been defined variously since the FCC first required "five minute" access in 1972, but all definitions have several elements in common: Channel or channels allocated to access use on a first come, first served non-discriminatory, non-commercial basis; no charge assigned for channel use; equipment and facilities provided free for public use; no content control over the access channels exercised by the cable operator; and that the user is responsible for the program content.

Access is usually categorized into public, government, and educational access, and specific channels are allocated to each, although in smaller systems they may be combined. Public access operations often provide free training in equipment use and assistance with production and publicity. Local government and educational institutions also may receive assistance in production and use of equipment with or without charge.

The access we enjoy today and hope for in the future is the result of the dedicated efforts of a small group who, in the late 1960s, saw its prospects. One source of their inspiration was the Challenge for Change program of the National Film Board of Canada. George Stoney, who had served as Executive Director of the Challenge for Change program, and Red Burns initially founded the Alternate Media Center at New York University as a means of encouraging access. The interns this program placed throughout the country were instrumental in providing the impetus for the access movement. Currently, a grass-roots organization, the National Federation of Local Cable Programmers, with almost 3,000 members, promotes the concept and reality of access.

This movement has reached the point where today in a community that has access requirements in its franchise, all that is required of a citizen to produce a program for his or her local access channel is the investment of personal time and energy. And, other than prohibitions against advertising, obscenity, libel, and lotteries, the individual enjoys the same First

Amendment rights on cable as the corner soap box orator does in a public park.

This is a truly astounding concept in today's culture and political-economic climate. No other medium has comparable practices. Although it is argued that citizens have unrestricted access to print, there is a cost to printing and distributing a message to the public that may far exceed that of public access on cable TV.

In essence, by accepting the concept of public access in the cable franchise, the public has declared that it is willing to subsidize access. This declaration is comparable to that of subsidizing our other major civilizing institutions—the library system, the park system, and the education system. While these systems are tax supported, the source of funding for access can come from a franchise fee, collected by the local government, the cable operator, the subscriber, or any combination thereof.

Access, then, is fundamentally a means for the citizens, the local government, and the local educational system to communicate with one another. It provides a means to utilize the most universal medium of communication within our cities, satisfies our basic democratic belief that individual opinion is important, and becomes a valuable part of a city's infrastructure.

The task in refranchising is to bring all this to pass.

FILLING A COMMUNICATIONS NEED

Access, while "free of charge" to the user and the viewer, is not really free: Not only is access a cost of marketplace entry but each citizen who subscribes to the cable system in his or her community through that subscription is paying in part for access. In today's economic and political climate, any city considering access requirements in its refranchising process cannot and should not ignore this. Every institution and public service today is facing budget cuts and must justify its value. What is the value of access?

It's overall value is impossible to quantify, but there are specific services it provides to a community that are amply described in many publications and throughout this chapter that can be accounted for. It is possible to acquire statistics of use, e.g., number of hours of programs produced and the diversity of groups utilizing the channels. But its real value to a community lies in the change it can make in the quality of communication within groups in the community and between groups and individuals. The need for this kind of connection between citizens is obvious in our cities beset by factions and the myriad differences that characterize each community.

On the "practical" side, what may be the most attractive feature of cable to the city in the long run is the range of information delivery services it can provide. Today, community bulletin boards, documentaries on local events or local institutions, community newletters, phone-in dialogue programs on local issues, public service announcements, time tables, bus schedules, and map information are examples of the types of services that can be cablecast over the access channels. Library catalog systems and other data storage and transmission systems can be facilitated by access cable. Of great value to many institutions and civic units is the possibility of producing and distributing training tapes and other information of internal value to the organization. Undoubtedly other uses will be found for access that will further enhance the value of the cable access system to the city's infrastructure.

Many of the services for larger cities will be found in the proposal for a separate institutional system; however, in small communities it may not be economically feasible to build a separate cable system. In these cities a scrambled governmental or educational channel on the city-wide system can provide the same internal communications possibilities. Additionally, a value can be placed on the public relations benefit that access provides local institutions and organizations. Cablecasting the town council and school board meetings makes public officials recognizable to the citizenry, and conversely, more accountable to the constituency. Public issues of importance to a political body may be more easily obtainable from call-in surveys than from public hearings.

Accessibility to All

If access is to realize its full value to a community, then the access channels must be accessible to the whole community. In many communities, less than 50 percent of the homes passed subscribe to cable. There are many reasons for not subscribing, but a significant factor is that many families cannot afford the monthly charge. If cable is to be the major channel for information transfer in a community, then these families become the information poor. To solve this problem the concept of universal service has been put forth. Universal service means that access channels are provided without a monthly charge (although often there is an installation charge to all homes in the franchise area). In some franchise bids, off-the-air channels are also included as part of the universal service tier.

ACCESS VERSUS TELEVISION

A franchise that makes provisions for access must rest on a thorough understanding of the nature of cable in general and of access in particular. Because cable appears on the same television set in our homes as does broadcast television and because this is all that most of us have known, it is hard to accept the idea that cable is not television. On the surface this seems to be an absurd assertion. But until it is accepted, the full potential of cable and its value to the community will go unrealized.

An analogy may help to see the distinction. Think of television as we know it as a swimming pool and cable as the ocean. They both contain water, you can swim in both, but they are very different. The swimming pool like television is restricted to a very limited space; it is owned and the owner controls its temperature, chemistry, use, and access. The ocean is vast; no one claims ownership to it in its entirety, although there are points where access is controlled, and it is inhabited by a wide variety of co-existing (symbiotic) forms of life.

This is not to suggest that the franchising process is meant to design a way to control the ocean, but we do put up breakwaters, employ a coast guard, build lighthouses, and establish fishing regulations. The controls that we do impose only work to the degree that we understand the nature of the ocean. What is the nature of cable and access? How does it differ from conventional "television"?

Motivations and Expectations

To begin with, access is an amateur activity in all the manifold meanings of the word. Television is a professional activity in all the various meanings of that word.

Access producers, being amateur, are often unskilled, may be self-indulgent, may ignore the audience, and may initially have low technical standards. They produce low- or no-budget programming. On the other hand, access producers only produce programs because they want to, because they are concerned, because they care about what they do and believe that the program content is important. They do it because it is an activity that matters in their lives and in the life of the community. One advertisement said it all: "Reading [Pennsylvania] citizens are creating and producing their own programs for cable television *with the camera that goes with the people, where the story is.*"

Conventional television is produced out of a different motivation. It is a commercial enterprise and to exist must make a profit. To do so it programs to satisfy the needs of advertising. It measures its success by marketing surveys and audience ratings. Experience has shown that local programming does not attract high ratings. Legally, each station is responsible for the content of all its programs so they must assume control over programming. Individual stations and networks do attempt to make significant efforts to meet local needs, but with their operational and fiscal constraints, the effort and the resources devoted to it are severely limited.

The technical and quality standards of commercial television inevitably mean that its productions are very costly. As viewers, most of our experience has been with these high budget productions and we have learned to accept these standards as the norm. But to impose such standards on access would be disastrous. High budgets and access are incompatible. It does not follow, though, that high quality and low budgets are incompatible. Access must be judged by different standards, and in communities that have established access, viewers have learned to view access for the communications tool that it is and find the programming most satisfying.

Purpose of Programming

A second difference is that access is produced by members of the community for the community. Conventional television programming mostly comes from outside the community and is designed to appeal to a mass audience. The latter results in our knowing more about wars in the rest of the world than what's going on in the neighborhood next door. We need to know both.

The quality of a city depends on what the citizens know about their city and what they do about it. Information about local activities often isn't adequately covered in the media whether it be TV, radio, newspaper, or other outlets. Access cable in city after city is providing information about local events, issues, and people. It stimulates participation and informs actions.

Even a cursory look at the kind of programming offered on local cable channels will show that people with access now do have the opportunity to find out what is going on in their neighborhoods and communities in depth and detail that was not possible before. Even in small rural communities cable has provided a significant increase in the quality of local communication.

Tremealeau County, Wisconsin, has interconnected its eight small communities by cable. Its interactive system serves the county's senior citizens, its educational system, and provides local news and event coverage as well. In other communities large and small, access is serving city councils, local health programs, neighborhood associations, ethnic groups, religious organizations, and a variety of other organizations. They utilize access to communicate within their groups and with the community at large.

Value of the Production Process

This communications activity defines, in part, another significant difference between cable access and conventional television. In access, the production process can be as important to those who make the program as the final program is to the community. Its value lies in what the producers learn as they conduct the research and do the scripting, planning, organizing, shooting, and editing. All this effort to clearly articulate their statement has a profound effect on both the individuals involved and their organization. The group is forced by the necessity of production to clearly come to terms with what they want to say. Their statements become more thoughtful, the quality of discussion higher.

This contrasts with commercial television, in which the product is all important. It must meet "professional technical standards." Very expensive equipment and highly skilled production personnel are required to reach these standards.

Working in this way has precluded television from covering local events in depth. The one- or two-minute clip on the nightly news can't explain why neighborhood zone changes should or should not be approved. But the neighborhood association on its own access program can bring together all parties and spend the time that it takes to explore the issue. True, such programming will not gain a significant share of the mass audience, but it can make a significant impact on the decision-making process. It also makes an impact on those involved on both sides of the issue. A commercial television production on the issue would never have this impact.

Channel Capacity

Finally, the difference between cable access and television is the difference between scarcity and abundance. In some of the new franchises there are as many channels reserved for access as there are for over-the-air television channels. Beyond access, cable adds many additional channels of special services. Abundance means that we viewers can be served with programs that meet special needs and interests. For many of us that programming may deal with what goes on in our communities, be it a neighborhood concern or an activity that forms the basis of a community of interest, such as an ethnic group, a religious belief, or health issues. With the abundance of access channels natural networks can form, allowing us to talk to one another across normal barriers of time and distance.

Cable will change our viewing habits in ways that cannot be totally foreseen. Just as browsing in a paperback bookstore is a popular activity, channel hopping in the hopes of finding an intriguing program has already become a pronounced behavior in cable viewers. Repeatability of programs and program series is a characteristic of cable quite different from broadcast television, meaning the viewer is no longer tied to the inconvenient, arbitrary schedule of the programmer.

With the interactive capabilities proposed in the new or larger franchises, more services will be available in the home . . . Education will be increasingly available in the home—truly a university without walls . . . With the advent of interfacing home computers to cable, some extremely localized data information bases may be developed. Cable's future potential may make it as important as the telephone, and who can imagine life without the telephone?

If cable becomes just more television as it now exists, then a great opportunity will have been lost. But if the promise of a different use of video is taken advantage of, if access does open up communication in new ways, then all can benefit.

It follows that because of scarcity and professionalism, television is necessarily very costly. Most access centers can operate for a year on the budget needed to produce one network commercial. Most small cities' annual budgets are less than the cost to run that spot. This kind of budget obviously influences the way television is used. The kind of thinking that goes into high budget operations is quite foreign to access, but it is the common way of thinking about television. If imposed on access, it can seriously distort what access is about.

ACCESS AND LOCAL ORIGINATION

Access and local origination share many of the same qualities and goals. Both are under the broad classification of community programming. Although similar, they are different in motivation and operation. The significant difference is that local origination (LO) is programming produced by and under the control of the local cable operator. The operator is responsible for content, the crew, and scheduling. Although LO productions can be very responsive and sensitive to community needs, they are produced under the same constraints that operate in commercial television—the LO channels are commercial. Advertising is sold, and it is presumed that eventually the channels will produce income for the operator.

LO programming is not necessarily all local. The channels are often dedicated to a theme such as Arts channel, Blacks channel, Health channel, and so on, and the cable operator usually plans to utilize satellite feeds to supplement the local production on these channels.

Local origination offers many advantages to community groups, the principal one being that they will not have to take the full responsibility for production that access will require. The LO producer can work with organizations, prepare the script, arrange for the production and scheduling, and assist with the publicity. For many organizations this will be preferable.

What the organization will give up is final control over content. Its members will not gain the benefits that often come from total responsibility for the production as described above. Also they will not have control over scheduling that access may offer. In sum, it will not be their program in the same way that an access production is their production.

In some cases these differences are meaningless; LO producers have proven to be fully cooperative and supportive. But to maintain this level of cooperation goes back to the franchise. The specific responsibilities of LO—the commitment of staff, facilities, and budget distinct from access—should be specified.

What cannot be specified is the cooperation between LO and access. Since the two do share common goals of community programming, depend on local groups, may even share channels, and can utilize the same volunteers and interns, it is imperative that a good working relationship and clear communications are developed so the two neither compete nor stumble over each other. This could cause considerable confusion in the community to the detriment of both.

SETTING THE FOUNDATION FOR REFRANCHISING

Regulatory Climate

The short history of cable regulation as it applies to access does not provide any comfort for a city in the refranchising process. Supreme Court decisions have overturned many of the regulatory policies of the Federal Communications Commission (FCC) and has deprived it of authority to regulate cable. Congress, as of this writing, is in the midst of attempting to rewrite the Communications Act of 1934 and to provide a basis for cable regulation. Until that effort is completed, regulation is a city responsibility.

The city must be well prepared for the refranchising process. The opportunity for refranchising occurs when the existing system is to be bought out or merged, or the franchise period has expired. In most systems there are many benefits to be gained for the community, such as increased channel capacity, interactive services, and access provisions. Because of the uncertainties at the federal level, several strategies are being developed in communities across the country to build into the franchise certain contractual agreements that must be adhered to despite new legislation or regulatory policies. The franchise can become a matter of contractual law rather than purely franchise regulation.

Tensions

Negotiations for a franchise are inevitably conducted in an atmosphere of overt and covert tensions. The various parties involved in the process have their individual or collective goals, needs, strengths, and weaknesses. How these tensions are dealth with will influence the final agreement on access. The parties are the cable operator, the city officials and city staff, the access organization, if there is one, the access users, and the public.

An obvious tension exists between the conflicts of the cable operator to maximize profits, the community's desire for low rates, and the need for funds to support access. Many municipalities argue that access is a service whose value to the community far exceeds the cost. In communities where access has had the opportunities to demonstrate its worth, it does have strong support.

Enlightened cable operators also see the value in access. A few operators feel that people subscribe to cable for the access package. Others that it enhances the overall cable package. But more to the point, if the access centers and activities are strongly supported by the cable operator, then the users become strong supporters of the operator and go to bat for the company in public matters.

A tension that can occur in the future happens when the realities of the implementation of the access package clash with the glowing expectations that are inevitably generated in the franchising or refranchising process. A big public, government, education access package becomes an attractive lure to snare the franchise. It also attracts community support for the bidding company. The offer should be taken seriously, but it must be looked at in relation to well-conducted needs assessment. An overly ambitious access package that the community is unable to put to use can quickly backfire on the community and discredit access in the long run. This does not mean that all the access package must be utilized immediately. Capacity should be reserved for future needs.

Some of the earlier failures of access stem from a mirror image of this. Inadequate access facilities, staff, and financial support were not utilized by the public in the early days of access and cable operators pointed at them as proof that access was not something that the public was interested in.

The most serious tension affecting access is that over obscenity. Obscene material is illegal on cable. State laws are generally very clear. The question becomes whether or not the access organization that controls the access channels has the power to exercise prior restraint. It seems that if people have heard anything about cable they have heard that it can bring X-rated programs into the home. In spite of the fact that this programming is not on public access, the spector of pornography haunts the entire access movement.

Although court decisions appear quite clear that prior restraint is a violation of an individual citizen's rights, there are factions who are unwilling to accept this. If the struggle erupts, it can be with devastating effects on access and can infect the whole cable franchise effort.

Needs Assessment

In refranchising, the community needs assessment is often different from those typically conducted in the initial franchising. In the initial franchise, the contenders usually conduct their own needs assessment to support their application for the franchise. Although it is a serious effort, it can be as much a means of generating community support as it is a way of finding out what the community needs for the cable system.

The needs assessment in refranchising can originate from the history of the existing franchise. If there has been any access or local origination program previously, those groups responsible can be contacted and surveyed as to what they have done, what problems they have had, what has worked for them, and what they would utilize in the future. If there is a citizens advisory body, its members are a source of evaluation of both past performance and future needs. The city's cable office or the city office responsible for the franchise should have records of problems and suggestions from citizens who have called City Hall. The cable operator also will have compiled records of citizen involvement that will serve as a source of input into the assessment.

This historical review will be useful; but if any significant changes are contemplated, then a fully-developed needs assessment should be conducted, separate from the bidding company's needs assessment.

If the community has not had access and only knows cable for its ability to offer a better picture and some of the satellite and premium services, then a needs assessment relevant to access planning should be accompanied by a serious public information program on what access has to offer. Access becomes a recognized community need after people understand what it can offer and, in some cases, after it has had the opportunity to demonstrate its value. This can be ascertained by surveying similar communities that have had ongoing access programs.

PROVIDING FOR ACCESS IN THE FRANCHISE

What makes access work? It starts with the franchise. Each city has its own unique characteristics, needs, and expectations for cable and access that the franchise should express, but each franchise and associated ordinances have to deal with a number of common elements. (Although the franchise is an agreement between the city and the cable company, the city ordinance that approves the franchise should also have attached an ordinance that deals with access specifically to set up the ancillary organizations and structures that will be associated with access. They cannot be separated.)

The franchise and associated ordinances should set forth the public policy that clearly deals with the following:

1. *Governance.* There have been four types of organizations that have assumed the responsibility for public access operation. First is the Independent Citizens organization. This kind of group originated expressly to operate access, largely in smaller communities with minimum access facilities or none at all. This type of access management structure is usually a non-profit group, member supported, which (1) forms on its own initiative; (2) develops a working relationship with the cable operator; and (3) sets up access facilities, policies, and some sort of financial base, depending on the locality. More often than not this type was found in communities cabled early.

A second type of organization is based on the constituency. In this case a local institution assumes responsibility for an access center. Libraries, local government, schools, and churches account for the umbrella institution. The host institution may or may not be responsible for the funding of operation, but participates in it. Or the institution may only serve as an umbrella and the access operation becomes autonomous.

A third type is access operated by the cable operator. The cable operator sets up the operation, staffs it, and opens its doors to the public. The public may have an advisory body, but with this type of access the operator has control of channels, equipment, and facilities, but does not control program content.

A fourth, and most recently developed, form of access management structure is the Access Management Corporation. This is a corporation set up by city ordinance to handle access. It is granted a portion of the franchise fee, and its operation and relationship with both the city and the cable operator is established by city ordinance.

2. *Ownership.* Who actually owns the access facilities? Depending on the type of governance, ownership and responsibility for access facilities can reside with the cable company, the access organization, or the umbrella institution. In most cases, this is very clear, but in cases where the cable operator provides equipment and facilities, ownership questions can become controversial.

Usually maintenance and ownership are hand in hand, but these also can be mixed; the responsibilities should be clearly defined in the franchise.

3. *Financing.* A stable financial base is critical to access as it is to any activity. Funds for the staff and operation of access have come from a variety of sources and should be stipulated in either the franchise, city ordinance, or contract. Sources of funds may be cable operator, franchise fee, institutional support, public support, gifts, grants and foundation support, and public fund-raising activities. A diverse funding base is generally desirable.

4. *Facilities and equipment.* A clear statement is needed of who is responsible for providing facilities and who is responsible for replacement and expansion. Again this can be a mixed source of cable operator, access organization, institutions, and others.

5. *Training.* Public access is meaningless without training. A clear statement of responsibility for providing training and what that training consists of is necessary in the agreements for the franchise.

6. *Appeal, review, and enforcement.* A mechanism for any of the concerned parties—cable operator, city, institutions, and citizens—should be set up so that problems, complaints, and non-compliance with the terms of the franchise can be settled with minimal disruption. Such entities as a regulatory commission, municipal cable office, or other entity to deal with this issue can be established. In any case the local government becomes the body to resolve such disputes.

7. *Channel allocation.* The franchise should specify the channels allocated to public, government, and educational access and should set forth a mechanism to increase or decrease this allotment based on utilization. It should be noted that full development of access will take time. Citizens are not aware of its potential and few initially will have the skill to utilize it. Our entire educational system is set up to teach citizens to use print; it is only beginning to include video literacy. Cities should look at access channels in the way some visionaries looked at park land a hundred years ago. They could imagine then how needed such land is now. As the

information or communication age comes into maturity, the access channels can become one of the more valuable city resources of the future.

MAKING ACCESS WORK AFTER THE FRANCHISE

The franchise is but a start. If well done it can provide a good base, but implementation perhaps presents as many problems as the franchise itself.

All too often those involved in the franchise look back with satisfaction at a job well done and then walk away assuming that everything is taken care of. Those who now come in to put the contract into effect do not have the strong foundation that would come from full experience with the franchise process. This would apply particularly when the new franchise is significantly different from the previous one, such as, including a new form of governance for the access operation or establishing a new production facility and staff.

Similarly, cable operators often send a franchise team to make contacts, conduct the needs assessment, and write the franchise. When the franchise is granted they move on and the operations team comes in. It takes a while for all concerned to establish new relationships and to learn the contractual obligations already specified.

Although a newly refranchised city already has a base to operate from and may have active citizen support, the new structures set up in the agreement still need to solve all the problems associated with a new operation.

For access to work, it requires that the citizens, the cable company, and the city all cooperate in good spirit. The many decisions concerning technical points, unclear areas in the franchise and the contract, and new developments in equipment unforeseen in the franchise, require a flexibility and cooperation on the part of all parties. At this point the value of a well-defined and broadly-supported consensus on the community's needs and goals becomes very apparent. It provides the means to arrive at acceptable changes and new relationships.

In the new franchises where a board of directors for a non-profit corporation has the responsibility for access, careful consideration should be given to the board so that the community perceives the board as being made up of those: who have credibility in the community at large; who do indeed represent the various elements of the city; and, above all, who have had experience in working as members of a board and who know the board process of conducting business, setting policies and priorities, and generating community support. Board members should be willing to devote the time and energy needed.

The Board should be appointed at the earliest possible time, preferably at the conclusion of the franchising process so they can be involved in the contract negotiations.

The Board must be perceived as and be in fact, a fully independent body separate from both city political control and cable-operator economic control. Individuals selected for any other reason, can seriously hamper development and put access in a bad light in the community.

Regardless of whether or not board members know about cable and access, the board, jointly with city officials with cable responsibility, the cable company, and interested citizens, should undergo a training and orientation period. The goal of these sessions should be to provide all concerned with a common base of knowledge regarding access history, practices, policies, theory, operations, and support organizations.

One of the first responsibilities of the board will be to develop a policy manual for the operation of access. This document will define conditions of use of the facilities and equipment; policies regarding training, copyright, non-commercial or not-for-profit limitations on productions; legal procedures; and procedures for handling complaints.

Recruiting staff and establishing personnel policies are yet another immediate priority. Staff selection is crucial to the success of access, and proper selections should be rooted in the special nature of access as described above. An access center is not a television studio and access programming is not television. All staff members must know and feel this in their bones. The staff is the primary contact with the community. They must first be skilled community workers who secondarily know the techniques of video production. They must genuinely like to work with amateurs and not be frustrated by the ineptness of beginners, be totally at ease with individuals of vastly different opinions, and be able to recognize and encourage all who come through the doors.

Access centers depend on volunteers to produce programs, so both staff and board must be apt in dealing with, recruiting, and maintaining the enthusiastic support of volunteers. In addition to providing the atmosphere that makes volunteer effort satisfying (and this ranges from the provisions for coffee to making sure equipment is well maintained) some mechanism for recognizing their efforts is important to maintaining volunteer morale. This group of individuals is the access center's most valuable resource and every effort must be made to nourish it.

The access board and staff must make special efforts to insure that the core of volunteers retains the diversity of the community. If this core group of volunteers becomes one that holds a single position, be it political, social, or cultural, then it may tend to either cause the development of antagonistic factions or cause segments of the community to reject access because it is monopolized by an "access elite." In either case access can lose the broad base of community support that it must have to retain its integrity.

Funding for access is as critical as for any civic endeavor. A balanced, reasonably-secure funding base is the board's responsibility, although the base for it must be the franchise. The most successful centers usually have a balanced source of funds from a portion of the franchise fee, from the cable company, and from the community. The budgeting process should encourage public input and be responive to the public shift in priorities.

The access operation and the cable community operations are so intertwined that a good, nonantagonistic working relationship based on a carefully phrased franchise agreement is essential to making access work. This takes good lines of communication on all levels from the technical and engineering to management. The two operations must mesh smoothly. In some cases, the cable operator has representation on the access board on an official or advisory basis.

Finally, access needs to be well managed. Nothing can destroy it faster than the equipment not working, schedules not kept, tapes lost or misfiled, and the phone not answered; all the significant and trivial aspects of day-to-day operation. Careful attention in access is required as it is in any business or service organization.

William O. Grant

7. Serving the Rural Areas: A Technical Review

The mere existence of over 4,000 CATV systems seems to establish the economic viability of cable service. A more thoughtful examination of these 4,000 systems however, qualifies the above statement, that cable is economically viable for reasonably sized towns and urban areas only. In smaller communities, and their surrounding rural areas, the economics of residence densities of 15, 10, or fewer homes per cable plant mile is severely inhibiting. This is evidenced by the fact that the entreprenurial CATV operator is conspicuous by his absence in these markets. In short, there is no cable TV, and little promise of any in the near future, in these rural areas.

The inherent inequity of this situation is readily apparent if we consider the purchase of a television receiver by both an urban and a rural resident. These are identical appliances and the personal investment is relatively similar for both parties. The urban resident, either through a proliferation of television broadcasting stations in the area, or through the availability of a cable television service, can select from, and enjoy, perhaps 5 to 10, and as many as 40 or more different channels. The country cousin on the other

---

Willam O. Grant is currently a Communications Specialist for the Rural Electrification Administration, U.S. Department of Agriculture. Previously, he served as a Transmission Engineer for N.Y. Telephone and as Applications Engineer and Product Manager for Jerrold Electronics. In total he has been a cable television engineer for the last twenty-five years.

hand, would be considered fortunate if he receives three clear television pictures, and this often does not even include the three major networks.

The inhospitable economics of extended rural areas with widely dispersed population are also apparent in many other aspects of rural life, for example, the availability of public transportation and police and fire protection services. Even educational opportunities at the secondary school level are more limited.

These inequities "come with the territory" perhaps, and to some extent, may always exist. Indeed, one can argue convincingly that there are compensations in the "quality of life" in the rural environment. The fact remains, however, that as the "information age" is thrust upon us, the probability of an ever widening gap between the rural and urban populations is very real and cannot be ignored. The merit, or demerit, of pervasive, high density exposure to television is a question for debate by sociologists. The potentially divisive impact of gross deficiencies between major population groups does imply an isolation of one from the other. If it is not economically practical to achieve parity, there would seem to be sound reasons to at least narrow the difference.

Rural electric and telephone development offers some interesting parallels to cable television growth in rural areas. Faced with equally inhospitable economic factors, a limited revenue base, and a large service area, it was largely enlightened federal funding and diligent applications of practical engineering, that so dramatically improved the lot of the rural citizen. Today, due to those programs, electric and telephone service is commonplace throughout the nation's heartland. It seems that the same philosophy and similar approaches may produce the same result for rural television service, or at least permit some improvement in the current imbalance of service.

There is, of course, a great deal of statistical data reflecting many hundreds of rural power and telephone operations. Diligent study of this data has produced a reasonably accurate rural profile for purposes of this discussion. It can be shown that on a national average, the density of rural residences is on the order of 5.5 homes per mile. Over 96 percent of these homes are served by distribution systems that can be extended 100,000 feet (18.9 miles). The balance of rural residences are extremely isolated, and random in nature. These residences usually are served by other means. For example, ranches in the west often use radio for telephone service, and sometimes even generate their own electric power. The rural power and telephone experience has shown that system designs that provide achievable "reach" beyond 100,000 feet are somewhat academic and will cost-penalize unduly the vast majority of rural residents.

The unquestioned success of the earlier rural electric and telephone programs was not produced by startling technological breakthroughs. In each case, there was an in-place, flourishing technology in the towns and cities. What made rural service feasible was an in-depth understanding of this technology as utilized in urban markets, followed by careful adaptation or modification to satisfy the different demographics which the rural application presented.

URBAN VERSUS RURAL APPLICATIONS

The urban cable systems are limited in geographical area and are characterized by evenly distributed, relatively dense populations. Potential subscriber densities on the order of 40 homes or more per cable mile are typical. Such density provides an attractive base of subscription revenues for cable television systems to support both initial capital investments and annual operating costs.

The rural application of cable requires covering large geographic areas with a sparse population base, e.g., 5.5 homes per mile or less. The fewer number of homes per mile offers a smaller potential revenue base and presents a less attractive, and in some cases a not even viable, profit potential.

Thus there are dramatic differences between rural and urban requirements. In some areas all applications below a finite level of population density, say 25 homes per mile of plant, have been considered economically impractical. This would be a natural tendency, particularly if the design techniques and construction costs as developed in existing urban CATV operations are accepted without question.

This article will explore technical modifications which may have a significant impact on system cost and will in turn impact on the economic viability of rural systems.

Urban Cable Technical Designs

The urban designer is faced with evenly distributed, heavy service drop tapping loads (e.g., drops to 40 or more homes per mile) throughout his entire service area. If system transmission signals can be maintained at high levels, the efficiency of the service drop taps (point where drops leave the feeder system to the home) can be improved, but these higher levels require higher output levels from all system amplifiers. (Amplifiers are utilized to boost signal strength and quality.) This can produce unacceptable transmission performance if the signals are passed through too many amplifiers connected in cascade.

In urban applications, a trunk sub-system is used to provide the backbone distribution of signals throughout the service area. Since many trunk amplifiers will be connected in cascade simply to reach the service area extremities, the technical acceptability of the signal must be considered.

In practice, the trunk plant is operated at lower, more conservative signal levels throughout so that the signal-to-noise ratio is acceptable at the system ends and the intermodulation distortion betters the acceptable range. In effect, the operating output levels are held quite low to reserve some rather high intermod distortion contribution from the second level sub-system, which is called the feeder plant.

The trunk amplifiers then are operated at conservative gain figures (22 or 24 dB typically) and no subscriber taps are inserted into the trunk cable. Because of the low gain, the physical spacing is relatively close, and often higher cost, lower loss cable is used in the trunk to further reduce the total number of amplifiers required.

The feeder sub-system demands frequent tapping for service drops, and to improve the tapping efficiency, is operated at substantially higher transmission levels. This requires somewhat higher gain in each amplifier, and imposes significantly higher intermod distortion contributions from each feeder amplifier. To limit to an acceptable degree this higher intermod contribution, the feeder sub-system imposes a limitation on the number of amplifiers that may be connected in cascade. This higher distortion is tolerable in overall system performance because of the conservative, intentionally-limited intermod the trunk sub-system introduced.

Limiting the cascade of feeder amplifiers means that additional trunk plant may be required more often throughout the area. The practice of not allowing taps in the trunk cable means that a parallel feeder cable must be placed to accommodate taps along trunk cable routes.

To summarize, the overall system performance then is the composite performance of low signal level, low intermod distortion trunk followed by high signal level, and high intermod distortion feeder. Most of the noise in overall transmission will be produced in the trunk system because of the many amplifiers in cascade operating at relatively low input signal levels. Feeder plant is usually less sophisticated since it is distinctly limited in cascade and does not require automatic level or slope corrective amplifiers, and most generally, since signal levels are high, lower cost, higher loss cable is used also.

If the ratio of feeder plant to trunk cable is high (typically four to one or better) and the feeder plant is less expensive, then across a large urban system, a significant cost economy is possible. However, this is conditional on the majority of the feeder plant being "off-route" and not simply paralleling the trunk route itself. Cost savings also depends upon a significant cost differential between trunk and feeder type amplifiers.

RURAL CABLE TECHNICAL DESIGNS

In the majority of rural applications, the probability of developing advantageous trunk-to-feeder ratios is very remote indeed. Since the system extensions in the rural areas will be quite long (100 Kft in the profile), the trunk would have to be extended practically the entire distance. The feeder cable used in urban designs is not particularly cost effective in rural areas. In effect, in the outlying, rural area, the trunk-to-feeder ratio will most generally be an unattractive one-to-one ratio.

In the "in town" segments of most rural applications, the situation is not much more promising. Several studies indicate that the towns involved will rarely require more than 10 or 15 miles of in-town plant, combined with 100 miles or so of rural plant. Even if subscriber densities in-town favored trunk plus feeder designs, the ratio overall would present a marginal cost reduction opportunity at best, and this would be offset substantially by the logistics of introducing several new units of equipment and requiring a greater variety in maintenance spares.

In the rural situation, the most cost effective approach would be a single cable, utilized both for signal distribution as the urban trunk cable is employed, and provision of subscriber drop access along its length as the urban feeder plant does. If this plant was operated somewhat above the low distortion signal levels of conventional urban trunk cable, but below the high distortion signal levels of the feeder sub-system, we might produce a more cost effective design overall. Quite obviously, the major economy available through this approach would be to eliminate a substantial amount of cable, and the associated placement costs.

Although introducing subscriber tap devices into the single cable introduces additional transmission losses, and consequently requires more frequent reamplification of signal levels, this penalty is accepted since a significant cost savings is produced as compared to placing two cables, using one for transmission distribution and the second for subscriber access tapping only. If a long trunk of finite length (say 25 amplifiers) were operated at conventional trunk design levels, at the end of the trunk the noise contribution which had accumulated would be very close to or at the overall system specified noise level. However, the intermod distortion contribution at that same point would be substantially better than the system specifications require. This is because, in the urban application, it is intended to extend off the trunk into feeder plant which introduces almost no noise but a very significant amount of intermod distortion. In the rural application, no large grouping of subscriber drops is found and while the trunk would be noise limited, it would not be intermod limited.

By using higher gain amplifiers and higher output levels, the amplifier spacing could be increased and even though the intermod distortion would increase, it would be at technically acceptable levels. In effect, amplifier gain is increased by redistributing the allowable intermod distortion along the entire trunk route rather than reserving some distortion to be contributed from feeder plant which will not be built. If the input levels are not changed, the noise contribution will not change, and overall system transmission performance at the end of the system would be identical with that produced through a trunk-plus-feeder transmission system. But the same number of amplifiers (say 25 units) would be spaced further apart, and overall system reach would be improved by 20 percent or more. This extended coverage imposes no cost penalty since the same number of amplifiers are used in either case.

The single cable design will require some limited placement of parallel cables, however. This is necessary since the amplifiers are spaced as far apart as possible to produce the lowest possible system cost. To do this, the operational amplifier input signal levels must be somewhat limited as a compromise between system reach, or attainable length, and economy in plant construction. It therefore becomes necessary to use parallel cables, as for some distance immediately preceding each amplifier, the cable does not provide high enough signal levels to serve subscriber drops adequately.

Assuming the minimum signal level to be delivered to any service drop is established as plus 10 dBmV, then any point along the cable route where the signal level is below plus 10 dBmV is not sufficient even if all of the signal energy were diverted to the drop. The tap device also introduces transmission loss to the service drop. The lowest value tap device available is an 8 dB unit. This tap, inserted into the main cable, will introduce 8 dB of loss to the service drop connection port. If the minimum level acceptable to feed the drop is plus 10 dBmV and the tap itself adds 8 dB, then we cannot satisfactorily tap the main cable at any point where the signal level falls below plus 18 dBmV. This produces a section of the trunk cable route identified as the "no tap" zone. In a system utilizing .500 inch cable, at 220 MHz transmission frequency, the no-tap zone would be approximately 800 feet. This may be reduced to 700 feet or so since service could be provided from a tap located at the plus 18 dBmV signal level point.

To accomodate service drops which might be required in these sections of a system a technique called "back feed cable" would be employed. This requires inserting a directional coupler in the main cable at the high level output of every amplifier which provides a convenient feed point with relatively high signal level. A second cable would be fed from this coupler parallel with the trunk cable back along the cable route to provide tap capability in the no-tap zone of the trunk. If there are no homes which need to be served in the no-tap zone section of the system, some economy is possible by simply not stringing the back feed cable. The system design, however, provides for the placement of the back feed cable, at any later date if necessary. Thus management may elect to place the parallel cable initially or postpone such costs until actual service requirements are encountered.

An alternative solution to the back feed cable is available and may actually be more desirable in some portions of the system. Utilizing this technique the amplifier would be relocated back along the cable route to where the output of the amplifier (a high-signal level) is acceptable for tap purposes. Obviously this may increase the number of amplifiers required if it were done frequently in a long system route. The addition of several amplifiers may also increase the end of the system noise and distortion but it would take several additional units before this penalty would be severe.

In the in-town portions of the system, however, there are persuasive arguments for reducing amplifier spacing rather than placing back feed cable. Except for two or three cable routes which might pass through town and be extended for significant distances out into the rural areas, the cascade of amplifiers in town will usually be quite small. In these cases the transmission penalty of close spacing units, even if an additional amplifier or two is required, will have little significance. The placement of several short sections of back feed cable on the other hand may well be more expensive than the additional amplifiers. Basically, the judgment in terms of system design techniques should be based on the economics of the specific area to be served. For this reason, as a general rule, at least in the in town portions of the system, back feed cable is likely to be uneconomical.

However, some parallel cable may be necessary in town. On the cable routes which extend into rural areas, the penalty for inserting many taps in town may be high. It is not the additional insertion loss of these taps that is the primary consideration here, since the number of taps inserted would rarely necessitate the introduction of an additional amplifier. The problem is the introduction into these town and rural cable routes of many more mechanical connectors due to heavy in-town tapping which may reduce overall system reliability. For this reason a second, paralleling cable would be used for the routes that must be extended any substantial distance after leaving town.

OTHER CONSIDERATIONS OF RURAL SYSTEMS

Following are some of the rural design considerations for development of the single cable system.

System Maintenance

The limited subscriber revenue base must support the initial capital investment as well as all long-term operating costs. In rural settings, it is unlikely that sophisticated engineering personnel will be available or affordable. Consequently, to provide rural cable service it is incumbent on the rural system designer to try to produce systems that can be operated over long periods with dramatically reduced levels of sophistication in both servicing and testing techniques.

For the above reasons, the single cable design incorporates a high level of automatic, self-regulation in all system amplifiers. This produces a system which is highly invulnerable to variations in signal level whether those changes are due to temperature variations or to infrequent or even incorrect equipment adjustments by operating personnel. Identical amplifiers would be used throughout the system, thereby reducing the logistic problems of maintenance spores. Additionally, all amplifiers would be operated at a single set of input/output levels, or at worst case, two sets of levels, which would reduce the complexity of long-term system operation. In urban environments, a lower level of amplifier self-regulation is sufficient and several equipment types are utilized because systems can afford and do provide more sophisticated maintenance programs. The rural system is not in such an enviable position and must take unusual steps to ensure low-level operating costs.

"In Town" Design Changes

Special conditions are also imposed in many rural systems by the central or core communities within those systems. As previously discussed, this is the "in town" area where residence density and tap density are at a higher level than in the purely rural areas. In many in-town situations, the single cable design should be modified so that amplifiers have higher input and output levels. This will increase the cross-modulation distortion contribution, but where the headend is nearby (and this is typically the case), this increase will be offset by the relatively small number of amplifiers connected in cascade. Raising the system operating levels (both input and output levels) requires no more amplifier gain, but will improve the efficiency of tapping the system in the areas where the tapping load is the most dense.

Rural Service Drops

Whether a single cable design or the more conventional trunk/feeder design is applied, there are problems associated with rural service drops themselves. In the rural environment it may be anticipated that some service drops will be unusually long as compared to the urban and in-town installations. Normally a plus 10 dBmV is adequate for the average 150 foot in-town drop, but in rural areas drops often run 400 or even 600 feet. This added drop length must be handled in a special manner to maintain adequate signal quality. Several possible solutions are available.

One solution is to use larger size cable drops; these cost more but introduce less transmission loss. In some cases, left over trunk cable from the construction of the system might be usefully employed as service drops. It is also possible, during the system layout process, to simply select a tap unit which will provide a higher signal input to the longer drops. This is quite practical but it does require that notation be made on the strand maps used for construction which depict where, and how long, these service drops are. The effect of "heavier" tap loading of this nature will shorten system reach somewhat if there are many long drops, but usually the problem can be accommodated in this manner. The rural service drop problem is not unique to the single cable design by any means, but is a special consideration in the rural application.

New Service Requirements

There may also be some confusion as to the ability of a single cable system to accommodate new service requirements which develop after the initial system construction has been completed. The transmission levels throughout the system are affected by the number of service points to which the system must feed signal. If in the future, additional feed points are expected, the rural system designer cannot utilize all of the available system gain initially but must reserve some capacity for such additions in a logical and reasonable manner. This condition is equally applicable to any cable system designer however, including the conventional trunk/feeder technique. For example, if 1 dB of reserve gain is provided in every amplifier section, a wide range of additional tapping is permitted either directly into the main cable or through back feed cable at almost any point along the entire system.

System Reliability

There is the question of system reliability when the single cable design is used since it can be argued that a long cascade of amplifiers must inevitably reduce the reliability. This cannot be denied, but it must be remembered that even a trunk/feeder design of equivalent length would require at least as many amplifiers in cascade.

There is also the question of impedance mismatches introduced into a single cable system by the many passive devices being inserted into the main cable itself. It is theoretically possible that reflections from many such irregularities will eventually degrade the quality of the television pictures transported through the system. This is a legitimate concern, but studies indicate that no visible degradation should occur until the number of devices involved is on the order of a hundred or more. A large portion of rural systems can be constructed using the cost-saving single cable design before this problem of reflections will become limiting, and in many systems, the limitations will never be reached.

The lack of actual operating experience with such systems, and the lack of measured performance data, does suggest that some research and further investigation of all questions would be useful. The Rural Electrification Administration (REA), an agency within the United States Department of Agriculture (USDA), has conducted a series of comprehensive tests on a rural system built to the single cable design concept. It is beyond the scope of this paper to duplicate this report or to quote from it at length but the results were all very promising and it is a fair statement that no technical disqualification of the design concept was developed. Interested readers may obtain a full copy of these tests, and the results obtained, from the Society of Cable Television Engineers (SCTE), 1900 L Street, Washington, D.C. 20036, for a nominal fee.

FRANCHISING THE RURAL SYSTEM

Cost Analysis of Cable Plant

Cost is the primary factor in determining whether rural areas can be served by cable television. Following is an analysis of the economics of the single cable design. While cost figures vary due to location, inflation, cost of labor, etc., average cost figures are used in this analysis. Each individual cost item is identified so that the reader may simply adjust figures for ones more appropriate to the specific situation.

The cost figures are based upon ten miles of direct burial underground construction under all three tap loading conditions (4, 8, and 12 taps per mile). The estimates provided are for a system with an upper transmission frequency limit of 220 MHz using amplifiers capable of subsequent retrofit for two-way service but initially equipped for one direction od transmission only.

Cost Estimate—Unit Cost (1982)

Item	*Unit Cost*
.500 Coax Cable (unarmored but jacketed for burial)	$ 220 per Kft.
Amplifier (includes AGC and ASC)	$ 500 each
A.C. Power supply	$ 350 each
Subscriber Tap	$ 15 each
Couplers and Splitters	$ 45 each
Small Pedestal Housings	$ 50 each
Large Pedestal Housings	$ 75 each
Cable Placement and Splicing	$1,400 per mile
Engineering and Overhead	15%

Then the cost of 10 miles of plant serving 4 taps per cable mile was determined to be as follows:

Cost Estimate—10 miles—4 taps per mile

Item	*Unit Cost*	*Qty. Req'd.*	*Ext. Cost*
.500 Cable	$ 1,170/mi.	10 mi.	$ 11,700
Back Feed Cable	$ 1,170/mi.	3.4 mi.	$ 3,978
Amplifiers	$ 500 ea.	26	$ 13,000
A.C. Power Supply	$ 350 ea.	6	$ 2,100
Sub. Taps	$ 15 ea.	40	$ 600
Cplrs. & Splits.	$ 45 ea.	35	$ 1,575
Small Pedestals	$ 50 ea.	50	$ 2,500
Large Pedestals	$ 75 ea.	32	$ 2,400
Total Materials			$ 37,853

Cost Estimate—Materials and Labor (1982)

Total Materials	$ 37,853
Engr. and Overhead (15%)	$ 5,678
Placement and Splicing (Labor)	$ 14,000
Total Cost (Plant in place 10 mi.)	$ 57,531
Cost per Mile (4 taps per mile)	$ 5,753

This analysis includes only items that differ in cost from a trunk/feeder system. Therefore, service drop cables and associated hardware, headend, and program acquisition costs have been omitted.

To provide a conservative estimate, the purchase and placement costs for 700 feet of back feed cable in every amplifier section are included. As discussed previously, this is an optional procedure, but it is included in 100 percent of the amplifier sections to establish the highest cost of a single cable system. Additionally, the most expensive approach to tapping is taken and it is assumed that a pedestal housing is utilized at every tap location.

Consistent with the system design philosophy, the cost estimates include a directional coupler at every amplifier output. The unit costs applied in this analysis are shown on the facing page.

The figure to construct plant with 8 taps per mile was $6,150 and for 12 taps per mile, $6,596. These estimates represent a reasonably conservative cost for plant in place tapped to serve subscriber drops along its entire length.

Utilizing a different design model with .750 inch cable reduces the amplifier requirement for 4 taps per mile from 26 amplifiers (using .500 cable) to 18. It should be noted that the longer amplifier spacings also increase the length of required back feed cables. The estimate for 10 miles with 4 taps per mile (.750 cable) was $6,981 even though it required 8 fewer amplifiers. This cost differential is almost entirely due to the higher cost for the larger size cable.

As pointed out earlier, the tradeoff is fewer amplifiers against higher cable costs. This study shows a cost per mile of $5,753 (.500 cable) against $6,981 (.750 cable) which is approximately a 20 percent higher cost for the larger cable size. Similar cost studies have resulted in comparable findings. Previous studies have shown that raising the high frequency limit from 220 MHz to 300 MHz (which increases system capacity from 21 channels to 35 channels) imposes a cost penalty of approximately 15 to 17 percent. Quite obviously, in rural applications, conformity to the 50 to 100 channels offered in urban areas would not be economically feasible.

In half a dozen direct comparisons, where both a single cable design and a trunk/feeder design were applied to the same application, it was found that the single cable design was less expensive every time. As expected, the cost reduction was more significant in the out of town portions of each system which are approximately 25 to 35 percent less expensive. Even within the towns themselves, cost reduction on the order of 15 to 20 percent are realized.

Operating Costs

In rural applications, cable service is usually provided not only to rural areas, but also to small towns or population clusters. By cable standards, these are very small towns and most probably could not support a cable television service by themselves. For these areas there are the basic program acquisition costs which are non-usage sensitive. For example, it costs just as much for broadcast and satellite receiving capability if 100 or 1,000 or 5,000 residences are served. The obvious solution is to expand the subscriber base to reduce the cost per subscriber for these basic program acquisition facilities. In rural situations, however, any such expansion immediately introduces the problem of additional plant miles with only a marginal increase in additional homes passed. Unless a certain percentage of market penetration is assured (subscribers produced per number of homes passed), the economics are very discouraging, and to some extent self-defeating. For example, if service rates are high, to ensure viability at lower levels of market penetration, then the higher rates of themselves discourage subscription and obviously will limit market penetration. If lower rates are initiated but market penetration is not achieved, then the very existence of the service becomes questionable.

In a non-profit type structure, such as a cooperative, for example, the question might be clearly put to the membership on a sliding scale basis. For example, after system cost estimates have been refined, but prior to commencing actual construction, one might quite accurately ascertain what sustaining service rates must apply for various levels of subscriptions, and subscribers could sign up in advance thereby assuring the economic viability of the system.

In a profit-oriented private venture, however, one has no such assurance of public support. Subscription is entirely voluntary in each instance of course, but in a cooperative, the possibility of higher sustaining rates, if necessary, is more realistically attainable. This suggests the possibility, perhaps even the probability, of a new generation of cable television cooperative efforts similar to many rural power and telephone operations. Existing cooperative entities are considering entering the field.

Even then, the difficult question of cross subsidy must inevitably be addressed. Persuasive arguments against long rural cable routes, at high cost but producing relatively few subscribers, must be balanced against the wider base sharing of the basic, fixed costs such as program acquisition facilities. It is a difficult and complex question which cannot be resolved with simple solutions, but the same problems were encountered in the power and telephone development programs, and workable solutions were developed. As those rural services became more universally accepted and widely used, the economic viability became more assured.

Can the same evolution for rural television services be anticipated? Certainly access to the service itself is essential to habitualizing use, and the enhancement of the services, reasonably assured by the massive levels of research and development in urban CATV operations, would seem to promise well for sustained and higher rural market penetrations also.

The urban system is a much more attractive, lower risk CATV investment, but the question we are addressing is, does that make all rural CATV systems impractical or unsound as investments. Undoubtedly, it imposes more severe penalties for marginal performance, and places heavier emphasis on efficiency and cost reduction, both in initial construction and long term operation. But these conditions are equally applicable to most business ventures in the rural environment that have a limited revenue base.

The undeniable fact is that CATV service has shown itself to be a salable commodity in urban markets. With the cost reductions available through the single cable system design, it is probable that *someone,* to *some extent* at least, will inevitably enter the rural markets with this salable commodity.

THE REFRANCHISING RESPONSIBILITY

Those agencies or bodies involved in small community and rural cable franchising situations must clearly recognize the difficulties of viable cable operations in these markets. The cable industry has an image of high profitability which may, or may not, be deserved in urban situations, but which is grossly incorrect for the rural markets. The rural cable operator must achieve maximum market penetration as quickly as possible if he is to survive at all. The urban operator obviously covets maximum market penetration but he can survive, even be profitable perhaps, at some lower levels of success. Thus he can exist with perhaps 30 percent market penetration for some period of time until public acceptance or marketing experience develops an effective program to improve penetration. The rural system enjoys no such comfortable margins.

In areas where small population centers are coming up for renewal or refranchising, it will be important to consider adjacent rural expansion or new development. This can be done by the population center franchising authorities within their jurisdictional boundaries or by their cooperation with nearby authorities from towns or counties. Multijurisdictional processes may provide a larger potential subscriber base for the cable system. Authorities should develop an in-depth understanding of the technological modifications—such as use of the single cable design as a framework for rural applications—that could reduce system costs significantly and thereby encourage investment in systems to serve the rural areas.

If it is fair to state that the basic function of rural franchising bodies is to expand the range of local services, then these franchising authorities must establish and maintain a healthy environment for marginally viable new services such as rural cable television. It is unrealistic to expect rural cable television systems to provide either the variety or sophistication of high density urban markets at the onset. It is not unreasonable, however, to hope that an economically healthy rural service might be able to expand its initial offerings somewhat, after a sustaining base of subscribers has been established.

A good analogy might be made with eight or ten party line telephone service, which was for a long time, the only economically viable level of rural telephone service. But as rural residents became accustomed to telephone convenience, and market penetration was assured, the telephone service was upgraded to four and two party lines, and even private line service became economically possible. While the initial eight party service was clearly inferior to private line services provided in urban environments, it was, to the rural community, much better than no telephone service at all. More importantly, it was the only economically practical evolution to universal private line service. Cable television services, with all the new functions it may be able to provide, such as shop at home, fire alarms, etc., may require an introductory period to permit the same evolutionary process to take effect.

Patricia Watkins

8. Refranchising and Low Power Television

What's Low Power TV?

Low Power Television (LPTV) is the first new broadcast service authorized by the Federal Communications Commission (FCC) in twenty years. It is an extension of the translator television service which primarily developed as a way to bring TV service to very rural western communities. (Translators are believed to have been invented around the same time and place as cable—in Oregon in 1948.) LPTV is seen as a way to increase service to those people in rural areas, to serve small neighborhoods or specialized audiences in urban areas, to offer a low-cost entry mechanism for minorities in television, and to provide a medium for experimental uses of television.

Essentially LPTV stations will operate in the same manner as broadcast television stations that have been watched for the last thirty years ("full" powered stations whose signals are carried through the air with no special hookup required for their reception). But since LPTV transmitter power is so very limited, its coverage range and construction costs are much smaller than those of other television stations. (While a full power UHF station is limited to 5,000,000 watts ERP [effective radiated power], the maximum power a low power station might have would be approximately 50,000 watts ERP. Their broadcast ranges would be around 60 and 20 miles

---

Patricia Watkins is Director of Low Power Television Hotline of the National Federation of Local Cable Programmers. Formerly she was station manager of translator, K56 A U, Columbia, Missouri, and prior to that she served as Program Director of FM station KOPN, also in Columbia.

respectively. Construction costs for full power television stations are between one and three million dollars. LPTV construction—not including studio equipment—will be between $50,000 and $250,000.)

The FCC has made the operation of LPTV stations extremely easy by eliminating many of the technical and programming rules which apply to full powered stations. LPTV stations have no minimum daily operating hours, no program source restrictions (any combination of local, satellite, or primary station origination is allowed), no limitations on scrambled programming, no signal quality standards, and greatly reduced reporting and record-keeping requirements. The physical spacing required between translators has been reduced for low power, so many more LPTV stations will now be possible.

THE LOW POWER-CABLE INTERFACE

A number of interesting possibilities for application make LPTV important to consider during refranchising.

Universal Service

Some municipalities have considered using LPTV to deliver a universal program tier. By using one or more LPTV channels, public, government, or educational access (or any combination) could be delivered inexpensively to everyone within a five to twenty mile radius from the transmission site. Thus, LPTV could be used to reach those who cannot afford cable subscriptions and those in rural areas which will never be cabled. Additional unused space on the channel (sideband and vertical blanking space) can simultaneously be used for utility load management, traffic signal control, or delivery of teletext print services to the handicapped or homebound. A few cable companies have included LPTV applications for this universal service in their franchise bids.

Rural Service

Many rural educational systems are examining LPTV/Instructional Fixed Television Services (ITFS—a broadcast service that requires a decoder) combinations to share teaching staff, train students in broadcasting, and increase communications between school and community. The Eagle Bend, Clarissa, and Bertha-Hewitt school systems in Minnesota have been operating such a system since 1980. Their success has prompted the state to fund a study to identify appropriate sites for similar operations and aid those school systems in obtaining LPTV stations.

The first major operating LPTV systems are two which Alaska has been operating since the early 1970s and one which the Board of Cooperative Educational Services (BOCES) in upstate New York has been operating since the mid 1960s. All of these systems feature a combination of local and Public Broadcasting Service (PBS) programming and, of course, are serving very rural areas.

Some rural areas have been using several translators as a substitute for cable systems. This practice will expand with low power television, especially now that the LPTV rules will allow scrambled signals and pay programming systems (STV). In a rural area, LPTV can be the mechanism to provide the first PBS service available, along with a combination of public, governmental, and educational access. Several LPTV stations could provide a distant network signal, one of the superstations, and a scrambled movie channel, for instance.

Several cable access centers have already filed for LPTV stations to reach additional rural areas. Low power television can also be used to connect two distant cable access center studios in order to share programming. This would allow the access signals to reach rural, unserved audiences in between, as well.

Other Benefits

LPTV may provide the insulation between the access programmer and the cable operator that both desire. By placing all public or governmental access on an LPTV station and then requiring the cable operator to carry its signal, the cable operator is absolved of its libel responsibilities. Additionally, the access center is then not subject to the censorship of the cable operator or totally dependent on it for a delivery system.

Of course, if your municipality will not be cabled for several years or if the current franchise has inadequate access provisions and will not be renegotiated for many years, low power may be the answer to quite a few problems. For example,

- a delivery system for public, governmental, and educational access;
- a specialized access channel for that significant minority in the community; or
- a medium to link information with your sister city, five miles down the road, served by another cable company.

LPTV in the Franchise Document

If your community is not in need of any of the aforementioned options, remember three items which should be included in every new or rewritten franchise.

The first is to require that the cable system carry the signals of some or all of the LPTV stations which operate in the franchise area. Since the FCC has decided not to extend the mandatory cable carriage rules to the LPTV service, only local franchise requirements can guarantee that these valuable, local program services will be available to cable subscribers. Some options are to require that the cable system carry:

- all LPTV stations whose signal overlaps the franchise area;
- all LPTVs that offer local programming;
- LPTVs which are operated by non-profit organizations; or
- those LPTVs offering unscrambled, i.e., "free" programs.

In lieu of a requirement that all LPTV stations be carried (and perhaps, anticipating a time when the FCC might eliminate *all* cable carriage requirements) a mandatory installation of an "A/B switch" on each set is a good idea. This allows a cable subscriber to switch from the cable input to a TV antenna system and thus get reception of non-cable carried signals. A mandatory A/B switch will increase each viewer's choices and insure that local programming can be received by all residents. Installation of this switch is very inexpensive when performed during hookup.

Concurrent with this franchise provision should be a guarantee that each cable subscriber be fully informed of the value of any outdoor antenna system which the cable operator may be willing to dismantle during cable installation. Many cable subscribers have unwittingly given away reception systems worth hundreds of dollars and were forced to continue to subscribe to cable services if they could not afford to replace the outdoor antenna system. These antenna systems will be especially critical for reception of LPTV or full service signals not carried by the cable system.

OBTAINING A LOW POWER TELEVISION STATION

As a broadcast entity, an LPTV station must be licensed by the FCC. The total possible number of LPTV stations in an area varies inversely with the number of broadcast entities already existing there. Areas well served by TV now will have room only for very few LPTV stations, while rural areas with minimal current service could be served by many LPTV stations.

The Commission has temporarily discontinued ("frozen") acceptance of applications for large markets and will currently accept LPTV applications and license stations only in very rural areas (Tier I—areas more than 55 miles from the center of markets 1–212). The FCC will then move to areas more than 55 miles from the center of the top 100 markets (Tier II) and then to the major 100 markets (Tier III) last of all. The FCC is expected to reach Tiers II and III sometime between the summer and winter of 1983.

Because of limited channel availability for LPTV stations in large cities, and expectations of great profits there, competition is intense in those areas, with approximately fifteen applications already filed on each possible channel.

So the length of time and competition to overcome in obtaining an LPTV license will be insignificant in very rural areas and quite substantial in urban ones.

The commission has given preference in any competitive LPTV situation to applicants who own few other broadcast entities (and none within the same market) and to those who have more than fifty percent minority ownership. Unlike in other broadcast services, the Commission has reserved no channel space for or awarded preferences for non-commercial operation.

* * * * * * *

Low power television will become a substantial communications medium by 1985, and its existence should be recognized in any franchise process. LPTV is a tool which can solve many problems in an inadequate, existing cable system, and it can also fill in gaps which even the most sophisticated cable system may engender.

Joshua Noah Koenig

9. Protecting Consumer Privacy

Cable television is in many respects like many other subscription consumer services and, like many other services, providers of the service may establish practices which can have the effect of infringing upon the privacy of the consumer. The maintenance and resale of a subscription list is a common example of a practice with serious consumer privacy implications, but which is widespread across many different businesses. Other common, but serious, privacy questions can arise from the sharing of individual consumer purchase records and the networking of consumer credit information. More examples could be cited, but the point to be made here is that cable television service is not different from many other consumer relationships which currently exist under a wide variety of regulatory oversight.

THE SPECIAL NATURE OF CABLE PRIVACY

A major distinction of cable is that it is, for the most part, a service whose very provision flows directly from a government grant of franchise. Therefore, the authorizing government body often feels more directly

---

Joshua Noah Koenig is Director of National Public Radio's cable audio project in Washington, D.C. Formerly he served as Deputy Counsel to the New York State Commission on Cable Television, and prior to that he was on the staff of the Cable Bureau of the Federal Communications Commission.

responsible for the fairness of the business relationship established with the public than it would in many other cases. In fact, this connection to the grant of governmental operating authority, and the quasi-monopoly nature of the service, could impose a higher standard of consumer protection responsibility upon the local government than it would normally face with respect to other consumer services.

There is a more important reason, however, why cable television represents a legitimate concern for government oversight of consumer privacy issues. Unlike most other subscription services, *cable makes a physical intrusion into the homes and offices of its consumers.* This local presence is effectively *permanent, continuous, "live"* (in the sense that the communications wires are operational rather than merely static), and often *"active"* (in those increasing circumstances in which information from the consumer's end of the wire can be communicated back to the proprietor of the service or to other locations on the loop). For this reason cable can be seen as a special kind of threat to the individual's privacy, a potential form of "electronic surveillance" which deserves a special degree of care on the parts of the service provider and the authorizing government body.

This degree of concern is consistent with the treatment of other services which also intrude into the private space of individuals. The strict regulations and even criminal laws governing the provision of telephone services provide a clear example of this concern; modern judicial decisions have expanded the envelope of legal privacy to include the confines of a public telephone booth in the street. It now seems clear that privacy invasions or potential privacy intrusions are less disturbing to the degree that they are more "remote," and more disturbing to the extent that they are more immediate to the individual's person and personal premises.

The increased concern with privacy issues that arises from the more active and intrusive cable services means that government controls to protect consumers have become more important and are a more common subject of debate as newer cable services and more sophisticated cable systems are introduced to the public. In fact, in recent years the privacy issues most often and most seriously addressed by government policy-makers concerned with cable are those issues which arise from the very intrusive nature of cable in the consumer's premises.

FORMS OF PRIVACY PROTECTION

Regulatory controls on cable for privacy protection generally take the following forms:

- Limits on the *installation* of monitoring or interactive equipment; or, in some instances, requirements for the installation of privacy protection equipment.

- Limits on the *polling* of subscribers or the *surveying* of subscriber usage.
- Limits on the *monitoring* of subscriber terminals or consumer behavior.
- Requirements for the obtaining of subscriber *permission,* often in written form, for any of the actions or practices listed.
- Requirements for the giving of *notice,* often on a repeated or even frequent basis, warning the consumer of the existence of these actions or practices.
- Limits on the *disclosure* of subscriber information.

This last form of control, a limitation on disclosure of information, is similar to restrictions placed on many other consumer services, but takes on special importance in the field of cable services when the protected information is obtained by the active monitoring of on-premises equipment. However, even information as basic as a list of subscribers and their addresses can assume special status in the eyes of government regulators if a presumption is made that basic cable service is a form of local utility and that subscribers who sign up are not merely dealing with a commercial merchant but are obtaining basic communications services. Under this view, cable service would appear more as a right or necessity of local residents, rather than a merely optional entertainment. This would be consistent with the special legal status cable systems enjoy in many jurisdictions, such as zoning exemptions, utility tax treatment, condemnation power, and entry rights to apartment houses.

Additional controls often placed on consumer information, and generally applicable to cable subscribers, include the following:

- A right of access by the consumer to information about himself or herself.
- A right granted to the consumer to challenge erroneous information.
- An obligation on the part of the record holder to destroy all records older than a certain age or after a specific event (such as termination of the cable subscription).
- An obligation on the part of the record holder to protect private information against theft or unauthorized use.

Additional Areas of Concern

Privacy issues can also be raised by a number of other cable practices. For some years cable subscribers in various parts of the country have complained about the delivery of programming to their homes for which they have *not* subscribed. Some consumers consider this to be an unwarranted intrusion into their homes; a kind of electronic junk mail. This situation is unavoidable with programs on the basic service package (often including access channels), but it becomes especially disturbing when the unrequested program intrusion is a pay service or some part of such a service. On many cable systems with relatively simple pay program security a subscriber who does not choose a pay service may still receive the audio portion of that channel; this can be frustrating for parents trying to keep their children unexposed to some offensive cable program material. Some local governments have required that the subscriber's request for exclusion of program material be honored by the cable operator.

Theft of cable service is a common problem for cable operators and it has several disturbing implications for the privacy rights of subscribers. One area of concern arises from the efforts of cable operators to enforce the laws which prohibit theft of service. In an effort to compile an adequate record of evidence against an illegal cable user, the cable company can employ monitoring techniques which could be challenged as inappropriately intrusive. In such cases a wire tap warrant may be obtainable from the local court.

Another disturbing consequence of service theft is that the information carried on the cable may not be adequately secure to protect the privacy rights of the subjects of the information. From this perspective, the cable system may face some liability for invasions of privacy even when it takes no willful steps of its own to misuse subscriber information. This kind of problem is heightened when the cable system carries data transmissions for banks, security traders, and merchants. Unlike the telephone system, cable provides no separate loops which are individually routed to participating private parties; all cable transmissions, even private data, are available in some form everywhere on the system downstream from the headend. Thus, security of information content becomes a special problem for which the cable operator must be responsible.

GOVERNMENT ACTIVITIES

Federal Protections

No federal law or rule expressly provides for consumer privacy protection on cable, but the federal government has a number of general provisions to protect consumer privacy in communications services. For example, the federal Penal Law makes it a federal offense to wiretap or otherwise intercept unauthorized private communications (Title 18 U.S. Code §2510 et seq.). This law also provides for serious civil liability for invasions of privacy (§2520). Its application to "wire" communications is broad enough to include cable services. For this reason alone it is critically important that cable operators obtain subscriber authorization to intercept or disclose anything which could be described as a private communication.

The federal Communications Act itself contains a general provision to protect the privacy of radio and wire communications (47 U.S. Code §605). Under this section, those in the business of providing private wire communications are prohibited from disclosing the content of those communications to unauthorized recipients, or from making beneficial use of the information received. Traditionally, cable communications have not been considered to be private; cable has been treated much like broadcast in form and the interception and divulgence of broadcast messages is not prohibited by §605. However, modern cable services now involve many forms of private consumer transmissions which can be protected by the language of this law. Therefore, we can assume that the monitoring of upstream interactive responses from the subscriber's terminal is covered by §605 (especially in such sensitive areas as electronic banking or shopping). Moreover, this section may also apply to the mere monitoring of downstream-only delivery of entertainment services if these are customized to any degree (that is, if the programming delivered differs by subscriber option and mere cable subscription does not imply a fixed profile of program delivery). Thus, although cable companies may keep track of the programming and services bought by each subscriber, they may not be able to sell such information to unauthorized parties.

The cable television regulations of the Federal Communications Commission do not expressly provide for protection of consumer privacy, nor have they ever so provided. It is not likely that the FCC will consider adoption of such rules in the foreseeable future. However, various bills have been debated in the U.S. Congress recently concerning the regulation of cable television services and it is possible that the Congress will eventually pass a cable law which might include consumer privacy provisions. The eventual adoption of a comprehensive federal cable regulation statute may mean a significant curtailment of municipal franchise authority. However, local government power to impose some form of consumer privacy protection will probably never be removed, and federal privacy protections are not preemptive of other protective measures adopted locally.

Notwithstanding federal actions, and certainly for the time being, local governments should assume that consumer privacy can be best protected at the municipal and state levels. Existing protections provided by federal statute can be used effectively, but should not be considered adequate replacements for state and local provisions keyed directly at the provision of cable services.

Municipal Protections

A wide variety of local franchising municipalities have included privacy protection provisions in their cable service ordinances or franchise contracts. Many of these provisions are made specific to cable, even though the privacy questions at issue are generally relevant to consumer transactions. The tendency to place distinct privacy regulations on cable is strong because municipal governments often feel they have little authority or general responsibility over other forms of consumer communications or subscriptions services. However, some local governments have segregated privacy protection regulations and certain other consumer protection issues in local laws distinct from the cable franchise or ordinance but more generally applicable to all similar consumer communications services. This practice has logic to support it, but a more careful approach would be to include effective restrictions both in general regulations and within the language of negotiated cable franchise contracts.

State Level Protections

A number of state governments have adopted laws or regulations to protect consumer privacy on cable. As far back as 1973 the New York State Commission on Cable Television adopted a provision within its technical operating rules for cable systems which restricted the transmission of "class IV" (an FCC designation for two-way or interactive service) signals. Under the New York rule, such signals may not be transmitted without the subscriber's express written consent and all subscriber terminals capable of two-way transmissions must be designed to allow the subscriber to prevent the return signal. This rule further requires that written notice and operating instructions be given to all subscribers provided with two-way terminals (NYCRR subtitle R §596.3[e]).

Soon after the adoption of the New York rule, the State of Minnesota enacted a cable regulatory scheme which included a similar restriction on transmitting or monitoring two-way signals without the subscriber's express written permission.

Illinois passed a privacy act in 1981 which applies to the provision of all information and entertainment to households by electronic means. This law prohibits, unless the subscriber is informed and consents, unauthorized monitoring of service use, disclosure of subscriber lists, disclosure of individual viewing habits, or the use of home security scanning devices. Violations of this law can bring fines of up to $10,000.

In 1982 cable privacy protection legislation was passed in California and

Wisconsin. The California law relates to all subscribers of telecommunications services but expressly applies to cable systems. It prohibits monitoring subscriber use without written consent and it prohibits disclosure of any individually identifiable subscriber information. Violation of this law is a criminal misdemeanor threatening fines up to $3,000 and a year in jail. The law is effective as of January 1, 1983.

The Wisconsin law applies a state standard to municipal cable franchises, and became effective in April 1982. It requires cable technology to prevent unauthorized monitoring (each subscription connection must have such a device). It also restricts monitoring, disclosure of subscriber lists, and electronic surveys (unless done with consent and monthly notice). Violation of these restrictions could result in franchise revocation or fines up to $250,000.

The legislatures of several other states have considered bills to provide similar protection of privacy to cable subscribers. As this essay was being written in the fall of 1982, the proposed legislation in Maryland (S. 890 / H. 1498) included language which would require termination of records at the end of a subscription, prohibit broad blanket subscriber authorizations, prohibit forcing authorizations as a condition of subscription, and mandate that all authorizations be revocable by the subscriber.

The proposed Massachusetts laws (S. 370 and S. 421) relate a number of cable consumer issues and include privacy protection provisions similar to those mentioned above. In addition, these bills would also prohibit any electronic consumer polling conducted on a non-basic tier of cable service, require most consumer usage information to be compiled only in aggregate form, and require monthly notice to subscribers of individual records maintained.

A proposed law in Michigan (H. 5897) would create comprehensive regulation of cable under the state's Public Service Commission and would include prohibition of disclosure, without some form of consent, of subscriber lists or information on usage.

A Missouri bill (H.B. 1294) would make it a misdemeanor for a cable system to disclose subscriber lists.

A large number of cable related bills were considered recently in New York and several addressed consumer privacy. One of these (A. 9848) would amend a state statute to prohibit installation without prior written consent of a monitoring device. Others (S. 8765 / A. 11052) would prohibit disclosure of subscriber information without consent, but would allow collection of subscriber data and its aggregation and use, but with a requirement of confidentiality and of subscriber notice and access.

A proposed bill in Pennsylvania would require local cable franchise contracts to prohibit disclosure of consumer information.

A GROWING CHALLENGE

The protection of consumer privacy can only become a more critical responsibility of local governments and cable system operators as ever more sophisticated services are made available by cable. Through electronic banking and shopping information a great deal can be learned of an individual's lifestyle. Records of ticket sales and reservations for travel and out-of-home entertainment can be very revealing. Even home security systems can be used to keep track of the movements of individuals. Individual responses to opinion polls could be recorded and maintained for a profile of a household's attitudes and beliefs. Personal use of information media, such as electronic magazines, newsletters and newpapers, can be similarly illustrative of personal habits. Moreover, even the specific items used within such media, such as a particular feature or article, could be determined for each individual. Although similar personal records could be compiled today from the subscriber lists of other media, cable represents special challenges to privacy in the degree of detail it could possibly provide on individual behavior and in the fact that it can be used as a single composite transmission medium for a wide variety of information and service sources never before so easily measureable at a single point.

The Industry Response

Many of the cable operating companies are very sensitive to the dangers their new services pose to consumer privacy, and a number of these companies have prepared and adopted privacy protection codes of their own to help ensure against inappropriate intrusions upon their subscribers. Warner Amex Cable Corporation developed one of the first and most reassuring of these codes for its interactive Qube services introduced originally in the Columbus, Ohio, market. Naturally, it is in the cable companies' own business interests to protect their subscribers from privacy violations and to maintain a structure of operations which continually reassures the public regarding the safety and reasonableness of their products. Local governments should take care not to impose privacy protection rules which conflict with the privacy codes of the cable systems in ways that ultimately reduce protection to the consumers.

It should be kept in mind, however, that the appeal of using interactive monitoring of cable use will increase for such commercial applications as measuring specific advertising penetration and even the responsive effectiveness of individual ads. The business community at large will soon learn to use cable and all of its marvelous capabilities to individually deliver and measure messages and services. The hunger for such applications must be tempered by effective controls on unauthorized consumer privacy intrusions.

A Final Note of Caution

As described at the start of this discussion, the consumer privacy issues raised by cable are in many respects no different than those raised by many other media and consumer subscription services. Local governments should make an effort to resist attempts to provide extraordinary levels of consumer protection for cable without some basis of logic. Special privacy protections are called for in those instances in which cable represents a special area of concern because of its very nature, but not all aspects of cable service are equal in this respect and some regulatory discretion should be observed.

Joshua Noah Koenig

10. Leased Access Policy Issues

Perhaps no subject is more disturbing to the executives of the major cable television operating companies than the loss of effective ownership control of the transmission capacity of cable television systems to unrelated channel users and programmers. This subject is raised in the context of the lease of channels on cable systems and the imposition of commercial access rights by local governments in their cable franchises.

Services offered on cable television systems are playing an increasingly important part in the communications revolution so widely discussed in recent years. Cable delivery will certainly be a key component in bringing a broad variety of telecommunications services to a large majority of both residential and commercial users, who will come to expect to use this medium for ever more varied entertainment, ever more sophisticated information and educational sources, ever more comprehensive marketing of retail goods and other services, and ever more subtle and effective communication on personal, corporate, and community levels.

Joshua Noah Koenig is Director of National Public Radio's cable audio project in Washington, D.C. Formerly he served as Deputy Counsel to the New York State Commission on Cable Television, and prior to that he was on the staff of the Cable Bureau of the Federal Communications Commission.

WHAT IS "CHANNEL LEASE"?

To reasonably discuss the subject of channel lease it is necessary to distinguish several related or confusing subjects. Channel lease as used in this discussion has nothing directly to do with the "lease back" of an entire cable television system. The "lease back" of a system was a common form of operation some years ago, and involved the erection and maintenance of a cable plant by a telephone utility company for the rental by the operating cable service company. This form of operation became extremely rare after the Federal Communications Commission imposed strict restrictions on telephone companies in the 1960s.

In comparison, the channel lease arrangement considered here involves as a lessor a cable television operator who usually directly holds a franchise to use the streets and who builds and maintains the plant, programs at least some channels, and sells a program delivery service to the public. The lease channel arrangement allows a party other than the cable system operator to use some channel space for private purposes or to directly market some programs or services to those members of the public who are subscribers to the existing cable system.

It may be similarly helpful to keep in mind that channel lease is a concept that includes all forms of third-party access to the cable system's delivery capacity. Therefore, one could argue that traditional forms of public acess, and government and educational access, are properly defined as channel lease (albeit for minimal or no lease charge). In general, discussions of channel lease focus on commercial channel uses, and this will be our focus herein as well, although many of the issues raised are more broadly relevant.

It is also important to distinguish the classic "arms length" channel lease arrangement from an increasingly common situation in which a "joint venture" is undertaken between the channel user and the cable system operator (or the parent or sister corporation). This arrangement is often hidden and hard to identify, but its key element is the retention of ultimate channel control by the cable operator. Some cable companies are withholding access to cable channels upon a condition that they be "cut in" to some equity share or revenue split. In April 1982, at a seminar on cable franchising sponsored by the Paul Kagan organization, the Executive Vice President of American Television and Communications (A.T.C.), Mr. Michael McCrudden, plainly stated that the joint venture approach would be the only one considered (and even that very cautiously) by his company for the introduction of new services on its cable systems.

FORMS AND USERS OF CHANNEL LEASE

The leasing of cable channels to users and programmers other than the owner or operator of the wires has already developed significantly on many cable systems and has involved the following types of uses:

- the long-term dedicated lease of digital data transmission links between discrete locations for a private commercial purpose;
- the permanent lease of one or more video channels for offering "pay TV" entertainment options to existing subscribers of the basic cable system;
- the lease of special transmission capacity for use by electronic security services offering home fire, burglar, and medical crises alarms;
- the lease of temporary discrete advertising on an electronic bulletin board channel (sort of a video classified column);
- the lease of audio channel space for commercial or non-commercial "cable radio" stations;
- the temporary lease of channel time for political campaign presentations; and
- the lease or free use of video channels for local community programming presented by the local government, schools, or members of the general public as a sort of video soap box.

The channel lease arrangement can be expected to provide the basis for many of the more exciting new developments on cable TV in the near future, including the direct marketing of retail goods to subscribers, the use of cable TV lines for direct banking links by means of electronic funds transfer, and the temporary use of interactive cable for research on advertising effectiveness.

In approaching the regulation of cable channel lease arrangements, local governments should keep in mind the nature of the lease channel users. Although some local public, private, and institutional users will express demands for channel lease, increasingly the greatest demand for such access will be by major commercial entities. These may include:

- competing entertainment programmers (with programming supported by both advertising and pay subscription);
- banks;
- major merchants;
- newspapers;
- home security services;
- information data base services; and
- a variety of large corporations with private communications needs.

THE REACTION OF CABLE OPERATORS AND THEIR USE OF THE LEASED CHANNEL FORM

As the use of, and demand for, the leased channel arrangement increases, cable television companies will confront the threat of the loss of their own editorial control of channel use and the loss of near-monopolistic dominance of their subscribing audience as a lucrative market for further development. Sensitivity to this threat is already apparent throughout the cable industry and reaction to it can be observed in many subtle ways. One example of this is the reluctant offering of a second, rival, pay TV program service by cable systems already providing one such service (e.g., Home Box Office and Showtime), especially when the company owning the cable system also has an ownership interest in the original pay service. This *controlled competition* is obviously preferable to an unwelcome loss of channel control forced by a public hungry for variety. Keep in mind that pay TV services are now almost always offered to the public directly by the franchised operator of the cable system, which in effect buys the program at a wholesale price from the packaging and distribution company, such as Home Box Office. The cable operator would certainly rather not see unaffiliated programmers using his cable system to market programs or services directly to the subscribing public, and in direct competition with his own service.

Despite the potential threat of mandated leased channel access, some cable companies have already learned to use the channel lease arrangement to their own significant advantage, and in some cases to the detriment of local governments.

Some companies have been operating private line data transmission links for special customers such as banks and large corporations. Manhattan Cable TV in New York City has such a service, and systems like this have been employed to distribute private data services from a single source, such as stock market information distributors, to a large but discrete number of commercial customers.

The Manhattan experience shows how close these services are to traditional carrier functions. In 1976 the U.S. Court of Appeals in Washington, D.C., held that such services (non-voice, non-video, point-to-point and completely interstate) were outside the FCC's authority to regulate as either broadcast or wire related services (*National Association of Regulatory Utility Commissioners v. F.C.C.*). That court decided that such services were too much like traditional intrastate carrier functions. As a result, the New York Public Service Commission soon afterward demanded that Manhattan Cable file "telephone" carrier tariffs for these services; this decision is still in dispute.

Clearly, the telephone companies see these cable services as directly, and unfairly, competitive with existing telephone operations. Because of the technical differences, and more aggressive marketing, cable data transmission services are priced substantially below comparable telephone data links. Moreover, cable charges usually do not increase with distance, as do most telephone charges. The cable system distributes all data and other communications by first bringing the message to a system headend and then feeding it "downstream" to *all* points on the system. Thus, no message travels any shorter or longer distance than others. However, privacy and confidentiality may suffer by the cable system design. Security of data is maintained only by separating frequency assignments (or by time-division multiplexing).

A number of cable companies have developed leased channel mechanisms for the offering of their own pay TV services to subscribers. In this approach the "cable" company holds the franchise and builds and operates the transmission plant, but only some of the program services ("basic" services) are offered by the franchise; the optional "tiers" and "pay" services are offered by a separate (but often related) company, which leases channel capacity from the franchise. This arrangement can be a means of avoiding payment of state and local regulatory and franchise fees based on the "gross receipts" of the franchisee. In localities which regulate subscriber rates on a strict "rate-of-return" basis, such as the State of Connecticut, this mechanism can be used to continuously demonstrate the need for rate relief on the "basic" system.

The subscribing consumer of these services often is ignorant of the fact that such services are provided by different companies. The responsibility for delivery of services and billing adjustments is sometimes vague. Moreover, inappropriate tie-ins of service occur (in which one company may limit one of a customer's services because of a dispute between the customer and the sister company involving a different service). More important, the antitrust and anticompetitive issues arising from the lease of channel space for cable program offerings are potentially disturbing. Cable companies could force customers to "tie-in" packages of programs, including mixes of basic and pay services provided by two different sources. Non-related competitive programmers may be denied access to the same subscribers or may be offered leased channel space on terms much less attractive than those available to the original, related, programmer.

DANGERS OF A LACK OF LEASED ACCESS RIGHTS

It is relatively easy to conceive of the basic dangers to the public interest in the absence of a right to use cable television channels. The availability to cable subscribers of a broad diversity of programming should always have been a primary objective of franchising authorities and cable regulators. However, this goal may never be achieved in an effective form if cable service to the American public is dominated by an ever smaller number of cable operators who offer products that duplicate the limited range and content of broadcast television.

The concentration of ownership of cable systems adds daily to the danger of monopolization of media control. The FCC recognized this issue with the establishment of telephone/cable cross-ownership restrictions in the 1960s and television network and station/cable cross-ownership restrictions in the early 1970s. (In recent years the FCC has provided waiver standards for the telephone/cable cross-ownership restrictions in rural areas. The FCC has also on occasion waived the TV network/cable and even the local TV station/cable cross-ownership restrictions, and a general review of the need to continue these rules has been undertaken. However, the Commission has consistently reiterated its concern with the dangers of concentration of media ownership.) In any event, the public is beginning to realize that the cable is an increasingly vital communications link and should not be under the complete content control of a single commercial entity.

The lack of a right of channel access makes real the danger of censorship of cable programming by the operator of the wire. This could take the form of commercial market protection, political bias, avoidance of controversial subjects presented to a sensitive public audience, and even matters of taste in sexual and other subjects. Because of the pressure which can be brought to bear on the franchisee by the local government, restrictive channel use could be a form of government censorship once removed and could be applied to program content which is not violative of any laws but which is merely objectionable to persons with influence.

The control of channel use could be used to exert effective business preferences on a variety of local service providers. Limiting use of the cable facilities could effectively "make or break" local home security service companies, for example. Special electronic banking arrangements could provide vast advantages to some financial institutions over others (keep in mind that several cable companies are partially owned by or have close ownership arrangements with major financial organizations). Major retailers and even newspapers may be forced to vie with each other for the exclusive opportunity to stay in business with electronic marketing and delivery systems. And through this all, a strong argument could be made that the interests of the consuming, viewing, subscribers are not well served if access to them as a market is monopolized and restricted.

Another problem with the lack of a right to channel use is the general misconception by municipal officials regarding their own role in franchise award and administration. As cable service has moved away from simple community antenna functions, so have municipalities come to believe that they are now contracting for the delivery to local residents of a wide variety of program services. In effect, the franchise contract often looks like an arrangement for "software" delivery, rather than for the operation of a distribution plant. Unfortunately, such local governments may eventually find that they have no enforceable power to make the cable company provide any specific programs or type of programs, or even to charge a reasonable subscriber rate for program services (as distinguished from antenna or transmission services). Even consumer fraud in the area of service promotions will be hard to control with companies that have discretion to change their service packages at will (much like weekly magazines may now change content and format without approval from or liability to annual subscribers). One area of consumer protection which municipal governments could be addressing by franchise contract, a right to use cable channels, has been largely ignored or underutilized as local officials focus on the program offerings of franchise applicants.

For purposes of local oversight and protection, the *consumer* of cable services can not be presumed to be only the receiving subscriber; potential cable programmers, service providers, and channel users are also "consumers" of the cable service and should be beneficiaries of the cable franchise. The cable business, for which a franchise is granted, is no longer simply the provision of local community antenna service, but rather it consists of the sale of entertainment and other programming (from a variety of sources), and the provision of electronic transmission services for a wide variety of purposes. These are quite different functions, which the authorizing street use franchises often fail to recognize.

CONSTITUTIONAL ISSUES

The "First Amendment" rights of cable programmers have received much attention and discussion lately. These rights arguably include the right to "speak" and to "publish" without government restraint, and perhaps even to "peaceably assemble for the redress of grievances" in some electronically public way. Few now argue that these rights should not apply to the programmers of cable channels (including generally the cable system operators), but a serious question remains as to whether the cable operator, as a transmission vendor, should be able to deny program distribution to others in the community. Unlike the newspaper publisher, who owns the physical elements of his own medium, or even the broadcaster, who merely utilizes the inherent communicative properties of the common earth and air, the cable operator needs government permission to occupy the streets owned and controlled by local governments on behalf of the citizenry. Until this situation is changed the cable operators can never claim full and unfettered rights to do what they choose with the transmission medium allowed to them by government award.

Some advocates of the cable industry have argued that the imposition of any third-party use rights to cable channels is an illegal taking of property in violation of the Fifth Amendment of the U.S. Constitution. This might be true of an uncompensated taking of cable channels or facilities during the term of a franchise which included no such provisions originally. However, the imposition of access or channel lease provisions at the time of franchise award, or as a general governmental obligation on all new or renewed franchises, can not be viewed as the taking of anyone's existing property.

HOW TO PROTECT LEASED ACCESS

The most serious problems facing local governments in this area relate to the form and degree of leased channel control. Many options are available, but none appear to be wholly satisfactory. Some consideration must be given to the fact that cable services are increasingly in competition with somewhat similar programming and transmission services, which are usually beyond local government control or oversight. These rival services already generally enjoy the advantage of an absence of local franchise fees, and any additional restraints or obligations on the franchised cable systems should be imposed only after careful consideration of the consequences and ultimate regulatory goals. The still rare but increasing situation of multiple local cable system operations makes these concerns all the more complex; access obligations should be carefully applied to avoid an innappropriate favoring or impairment of one cable system.

The imposition of *common carrier* type obligations and restraints on cable systems is not unlawful, despite the arguments of some in the cable industry. The FCC has been prohibited from imposing such rules on cable by order of the U.S. Supreme Court (*FCC v. Midwest Video,* 1979), but this decision was based on the *statutory* restraints placed on that Commission by Congress and the very form of jurisdiction chosen by the Commission to regulate cable. It is true generally in the law that no provider of service which does not choose to act as a carrier can be forced to do so, but it is equally true that the government can simply prohibit anyone from providing a service through the use of public streets unless the service is provided in common carrier form. The validity of such a law would be tested on the basis of the service provider's right to use the public streets, which right has not yet been established for cable companies.

However,the decision to impose carrier type controls over cable should not be taken lightly. This form of regulation was designed to impose the most rigid form of *stability* over services whose exact nature was well defined and held to be highly valuable to the public. This degree of stability would have stifled the development of cable (or primitive community antenna TV) years ago and could yet deny many further public benefits. Moreover, any form of common carrier treatment raises awkward questions of government control of such critical factors as the rates charged for channel lease, the resolution of competing demands for scarce or particular channel space, and the competitive advantage enjoyed by the cable operator in self-dealing channel use on his own system. None of the various leased channel regulatory proposals debated to date has adequately addressed these issues.

Special Concerns

Local governments considering some form of leased channel obligations should also be prepared to deal with the following facts:

- that some indirect control of cable channel content will be lost to government officials and the community generally;
- that some forms of consumer protection oversight will be loosened for services which are not directly offered by the contracting franchisee;
- that some franchise fees may be lost as the gross revenues from some cable services are collected by parties other than the cable company;
- that the liability of the cable operating company for program content on leased channels must be limited in the most perfect way possible, and the resulting liability of the local government must be reexamined; and
- that the control over the conditions of service and the rates charged to subscribers will be weakened, even for basic services offered by the franchisee.

A PUBLIC OBLIGATION

Despite the difficulties of adopting a workable channel lease obligation, some efforts towards this end appear warranted. The issuance of a municipal street franchise for cable TV service is still usually an effectively exclusive award of public property. To allow the franchisee complete control of content is a possible abandonment of public trust. The government's responsibility to protect a "public forum" has been repeatedly recognized by the courts, and was so dramatically as recently as 1980 in the field of public TV broadcasting in a decision by a Texas federal district court (*Barnstone v. University of Houston*). It may still be assumed that there are practical limits to issuance of multiple cable franchises for municipal street use.

Local governments may very well decide that they have a responsibility to create and protect what is the electronic wire equivalent of a *public marketplace* or a *public highway* for communications. For hundreds of years it has been the government's recognized job to use public property to create and maintain highways and markets. The government need not, and probably should not, itself directly control who can use the highways or markets, or what is transmitted or sold thereon and therein, or at what price or to whom. Nor need the government, nor should it, abandon its responsibility by giving the public market or highway into the hands of a single private commercial owner with unlimited discretion to restrict or manipulate the use of the market or highway by others.

Appendix

Requesting Proposals for Cable Television Systems: A Model for Municipal Governments

Table of Contents

PREFACE

Requesting Proposals for Cable Television Systems: A Model for Municipal Governments was written by members of the National Federation of Local Cable Programmers (NFLCP) to provide local franchising officials with a sample structure they could follow when developing a Request for Proposals for cable service. The NFLCP is indebted to the Nordson Foundation, which funded the development of this material.

It is important for the reader to understand the purposes for which the model was developed. The NFLCP is vitally interested in the development of cable television as a medium for community communications. Local programming and cable services which improve the delivery of community services are often overlooked in the franchising process, when obtaining the more visible and glamorous services cable offers--such as uninterrupted movies and 24-hour sports--is the more immediate goal. The purpose of the model is to help community franchising officials know what to look for when they are planning to work in conjunction with the cable operator to see that the maximum potential of cable is realized and that a system which will be indispensable to the community will be properly built and maintained. The model is NOT intended to replace the services of a consultant or other expert who can analyze the needs of an individual community and develop an RFP which is appropriate for that particular community. In fact, we expect that the model might suggest how complex a process planning for cable is, and how necessary expert advice is.

The model we developed has many limitations which should be kept in mind by any reader. First, the model is not comprehensive. It is difficult to cover every area that should be included in a cable RFP, in part because building a cable system is such a complex process with many parts that will vary from community to community and company to company, and in part because the state of the art in cable is changing so quickly that no print publication can be current with all of the latest developments. Also, it is the nature of any model to be general in its description; this model, being no exception, cannot possibly take into account the configurations and needs of every community.

The model will not serve any community well if it is rigidly followed. Its best use would be as a suggestion of the depth of detail an RFP can specify and as a checklist of items for a community to CONSIDER including in its RFP. Common sense and expert advice can sort out what individual items are or are not useful for a particular community. As we stress in Section 1, cooperation is the key to a successful cable system. If franchising officials, cable operators, and cable users work together during the franchising, renegotiation, and refranchising stages, everyone can benefit from having a well-used community resource. We hope this model will help contribute toward that goal.

The following individuals contributed to this Model: James Bond, Sue Miller Buske, Brother Richard Emenecker, Dan Jones, Brian Owens, Diana Peck, Robert Pepper, Jean Rice, Jerry Richter, Thomas Sherard, and Don Smith.

1. PRINCIPLES OF FRANCHISING, REFRANCHISING, AND RENEWAL

Ten basic principles are at work in the cable television franchising, refranchising, and programming process. To realize cable television's full potential as a communication resource, every community must be aware that:

1. Every cabled community has the capability and responsibility of working with its cable system operator to develop local community communications services.
2. The community's first responsibility is to become knowledgeable about the technical capabilities and operational structures of a cable television system and the potential of that system as a community communications resource.
3. In the cable franchising and refranchising process, the local government and cable company are bargaining agents negotiating a contract. The local government's role is to define, require, and assure maximum cable services for its community. The cable company's role is to guarantee delivery of those services and to assure a reasonable return to its investors.
4. Community service provisions, especially those involving local programming, should be assigned highest priority in the negotiations.
5. It is solely the community's responsibility to determine the level of local services it requires or will accept.
6. If, during the negotiations, the community requires, or the cable company promises, too little or too much, neither's best interest will be served.
7. The community should require only those services and provisions for which it has established a need and for which it is able to guarantee support.
8. The cable company should offer to provide only those services it is able and willing to deliver.
9. The franchise contract between the community and the cable company should be opened for periodic review and renegotiation as new communication technologies and community capabilities emerge.
10. After the franchising process, the community, represented by the local franchising authority and a nonprofit Community Access Corporation, must form a partnership with the cable company to provide community programming services.

Principle #1

Every cabled community has the capability and responsibility of working with its cable system operator to develop local community communications services.

The traditional role of local governments in the franchising and refranchising process has been to require only that cable operators build and maintain adequate systems for the retransmission of television and radio broadcast services. Until recently local governments had not been involved in requiring and requesting cable operators to provide channel capacity, funding, and facilities for local community service uses. The FCC attempted to require such channels from the federal level, but was ultimately unsuccessful. Community service provisions are being required now in the franchising efforts in major metropolitan areas because of the growing awareness on the city's part that cable is a valuable communications resource, and that the city government is the only body empowered with the authority to require the cable operator to act in the public interest. In addition, there is growing awareness on the cable industry's part that providing adequate community programming capabilities is good business.

Principle #2

The community's first responsibility is to become knowledgeable about the technical capabilities and operational structures of a cable television system and the potential of that system as a community communications resource.

A. Technical Capabilities and Operational Structure: Cable television has been and still is a business devoted to distributing television signals to subscribers' homes. These signals are received through a central antenna, satellite earth receiver or telephone lines, and redistributed through a series of amplifiers over coaxial cable to subscriber converters and television receivers. The number of television channels available on these systems range from 12 through 108+. Most older systems carry 12 channels, more recent ones 20-35 channels, and most proposed systems 54 and up.

Service to subscribers are usually divided into two categories: basic and pay. Basic services usually include those broadcast and satellite signals that cost the cable operator little or nothing to acquire. Those signals are from local and distant network, independent, and public broadcast stations which in many cases can be received on a good home antenna. Distant broadcast signals from such stations as WGN of Chicago or WTBS of Atlanta are received through microwave systems or via satellite. Other satellite services the cable operator may currently choose to provide as part of the basic service include 24-hour cable news and sports, a children's channel, and live coverage of the Congress, as well as religious and instructional programs.

Local programming is also usually included in the basic package. The specific nature of these programming services will be discussed in Principle #4, but they include local origination programming provided by the cable operator, and community access programming provided by local institutions, agencies, organizations, and individuals. Some automated channels featuring local and national news and weather, financial reports, and community calendars may also be offered as basic services.

B. Cable Television as a Community Communications Resource: Communities around the country are gradually developing the capability of accessing and using cable television systems to meet local informational, educational, cultural, social, and entertainment needs. The largest metropolitan areas and smallest towns are discovering that cable television can be used to provide inter- and intra-community communications; that community-wide and neighborhood-specific issues and concerns can be explored through local programming efforts; that libraries and educational institutions can provide access to information sources and instructional programming via cable; that community cable programming centers offer the opportunity for individuals, agencies and organizations to develop and use communications skills through mastery of available telecommunications technologies; that these centers and their activities, such as live coverage of local government meetings and public forums, foster increased understanding of and participation in the democratic process; and that access to the cable communications system encourages an expression of and appreciation for individual and community artistic, intellectual and cultural heritages.

Cable television revenues also provide the funds that make this access possible. Application fees from bidding companies can be used to educate the community about the uses of cable, to design future program efforts, and to develop a Community Access Corporation. A percentage of the successful cable company's revenues should be used to provide on-going support for local community cable services as should a percentage of the franchise fees paid to the local government.

Cable television also develops human resources by providing training and jobs for systems managers, office personnel and technicians, for local program facilitators and producers, and for organizational and institutional communications specialists.

Pay services currently being offered on cable include the movies channels, special entertainment programming, some professional educational courses, and home security systems. Future common pay services will likely include data transmission, electronic mail and newspapers, meter monitoring, home banking, and shopping. In the near future, however, movies, sports, entertainment, and arts channels will remain the major pay services offered via cable.

To access and develop these community cable resources requires a major local and national educational effort. Local governments should organize cable advisory boards to identify information sources such as publications, cable consultants, state cable commissions, and national organizations whose goal is to assure access to and use of cable communications systems. Only by educating themselves, their local officials, and citizens will these advisory groups and the communities they represent be able to deal with the complex task of developing and maintaining access to cable resources through the franchising and refranchising process.

A note of caution:

In hiring outside consultants to educate your communities about cable, to define your telecommunications needs and capabilities, to develop and evaluate bid proposals, and to design local programming operations and facilities, be aware that you are hiring a point of view about the value of local community cable services. Not all cable consultants are aware of the contributions that community-operated access centers have made to towns and cities around the country, nor do they endorse or encourage direct community involvement in setting cable rates and requiring specific commuity programming services.

Principle #3

In the cable franchising and refranchising process, the local government and cable companies are bargaining agents negotiating a contract. The local government's role is to define, require, and assure maximum cable services for its community. The cable company's role is to guarantee delivery of those services and to assure a reasonable return to its investors.

This franchise contract is based on the granting of a privilege by the governing body. Applied to cable television, a municipality grants a cable operator the use of the valuable right-of-way of public streets for stringing and laying cable. The language of the contract will reflect the community's definition of its specific needs and expectations of the cable service provider. It will also include provisions guaranteeing that those needs and expectations will be met by the cable operator.

It is absolutely critical that the local government be totally responsible for the language in the franchise contract. Under no circumstances should a contract drafted or redrafted by the cable company be accepted by the franchising body as the final document. Here, as in most states of the franchising and operation of the cable system, local officials must identify individuals within the community with interest and expertise in cable television matters. They should also seek advice from experts outside the community. Once again, the expenses for these services should be paid by the bidding company or companies in both franchising and refranchising.

A note of caution:

Accept no good-faith offers from bidding company representatives before, during, or after the franchising process. Get everything in writing. In many cases, the company respresentative who is negotiating the contract will not be responsible for implementing it. The local management may not be aware of community service commitments or may assign them a lower priority. The original negotiators for the company may change jobs or move to other corporations. Local officials change as well. No one from the local government that negotiated an original franchise document may be around five or ten years into the contract. The franchise document may remain as the only reliable source of information about the terms of agreement for both the community and the cable company.

Principle #4

Community service provisions, especially those involving local programming, should be assigned highest priority in the negotiations.

In most franchising or refranchising negotiations, technical specifications, economic projections, construction schedules, damage deposits, and performance bonds are the major criteria upon which proposals are judged. Community service provisions, however, especially those involving local programming, are beginning to become more important in franchise negotiations.

In developing the RFP, local officials must clearly distinguish between two types of local programming to be provided to the community.

Types of Local Programming Operations:

One form of community programming is local origination (LO). LO programming is usually initiated and controlled by the cable company. It may include coverage of some community events such as city council meetings, but usually focuses on sporting events and other activities that might attract fairly large audiences and subsequently some advertiser support. LO channels are similar to small-scale commercial broadcast stations and usually depend upon advertising revenues for their existence. Their staff members may have other responsibilities within the cable operation, frequently marketing. As a result of these factors, many LO operations cease to provide community programming if advertising revenues or staff allocations are reduced. A marginally profitable cable company will not carry a local origination effort that is seen as a losing proposition.

The second type of local programming is called "access." Access programming is produced by community volunteers or local nonprofit organizations and institutions. The content of that programming is determined by the individual, group, or organization which desires a program and produces that program. Access programming is normally noncommercial in nature. Several types of community programming traditionally fall under the umbrella of access. Those are public, educational, arts, minority, and all other noncommercial community produced programs.

In many communities, especially those with limited channel capacities and financial resources, cable companies have asked libraries, high schools, colleges, and universities to provide local programming. Usually the cable company provides funding equipment and the institutions provide office and studio space, and some support staff. In Bloomington, Indiana, for example, the access center is housed in the Monroe County Public Library and is funded by the library, city, cable company, and outside grants. Similar systems have developed in Iowa City, Iowa; Ft. Wayne, Indiana; and Rome, Georgia. In Frankfort, Kentucky, a municipally-owned system provides the community channel; and the East Lansing, Michigan, system carries a full complement of access channels operated and supported by the cable company, the city, the library, the school system, and Michigan State University.

Other access centers are operated by independent nonprofit corporations supported by foundations and local grants. Berks Community Television in Reading, Pennsylvania, has been providing community programming since 1976 with support from the local business community, the city, and recently the cable company.

An emerging operational model in small communities is the community consortium. Under this structure, part of the funding and staff support is provided by those agencies and institutions who use the services of the access center to serve their special audiences. It makes no sense in most communities to duplicate expensive production, duplication, reception, and distribution facilities. Nor is it feasible to have separate administrative and operational staffs.

A community access consortium allows the local government, library, educational institutions, social service agencies, cultural, professional, and arts organizations, business industry, and the cable operator to develop and maintain viable community programming efforts. The consortium broadens the center's funding base, reducing the possibility that community communications services will disappear because of major changes in any one funding source.

The purpose of the consortium and its operational structure must be clear. The members not only have the opportunity to use the system, they have the obligation. And although one function of the center will be to deliver educational and informational programs, its primary purpose is to promote creative and constructive use of the local cable system and to assure that individuals and groups are taught to use it.

Another value of the consortium is that it pools human, program, financial, and technical resources. Most importantly it brings people together from various groups and institutions to create community-responsive programming. As a result they are able to identify common needs and concerns and to derive mutually beneficial solutions.

Specific community service provisions with realistic support levels are defined throughout this document. The most important point is that these provisions be given the weight they deserve. For the community, these services will be the local legacy left by the franchising officials. For the cable company, they will be significant investments in community relations and genuine expressions of community involvement.

Principle #5

It is solely the community's responsibility to determine the level of local services it requires or will accept.

Although outside consultants and the bidding cable companies should play major roles in helping communities define their telecommunications needs and in developing effective and innovative techniques for meeting those needs through cable, the community must be involved in that process from the initial planning and design stages throughout the construction and operations phases. Unless the community is directly involved in developing the specifications for the services, it will not be sure they are being provided and will certainly not be in a good position to use and support them. It is vital that representatives from the local government, informational and educational institutions, social service agencies, business and professional organizations, and arts and cultural groups--all potential users of the system--be directly involved in determining what the community needs and will use. If a bidding company offers a service or a facility that was not required or requested in the RFP, the franchising group should conduct a thorough study of the provision. It might not be something that the community desires or will use. It might even become a burden. It is a mistake for the community to rely too heavily on the advice of outside consultants or on the recommendations of the bidding company. When in doubt, get a second or third opinion on specific provisions.

Principle #6

If, during the negotiations, the community requires, or the cable company promises, too little or too much, neither's best interest will be served.

The competitive nature of the cable franchising and refranchising process and procedures has led to an escalation of demands on the part of the community and an escalation in promises on the part of the cable companies. In some cases, the community demands are prohibitively expensive, even though companies will agree to meet them. The result: delaying tactics by the companies in meeting the demands, confrontation court battles, more delays, ultimate loss of the facility or service, and a certain deterioration in the relationship between the cable company and the community. If the company makes excessive promises without being asked, the same results occur.

Most communities, both large and small, are relatively unsophisticated in cable matters and may inadvertently require too little of franchise or refranchise applicants. If the bidding competition is not keen, or there is no competition, communities will not be forced by competing applicants to learn about available cable system capabilities and possible services. Ideally, all bidding operators will offer state-of-the-art technology with realistic community programming facilities and funding in every community. Realistically, however, many communities have awarded franchises without doing any research into what minimal requirements should be and how they can guarantee in writing that the cable operator will carry out ideas proposed orally. When this occurs, not only will one community be locked into inadequate franchise provisions, but the cable company and local officials will lose credibility. The cable companies' national and local franchising and refranchising efforts will suffer. The local officials may experience political disfavor once the public learns about the inadequacies of the cable system in its community.

Principle #7

The community should require only those services and provisions for which it has established a need and for which it is able to guarantee support.

The emerging franchising and refranchising efforts in urban areas have been characterized by contractual provisions that are somewhat one-sided. The community grants the cable system use of its right-of-ways in return for the services and provisions outlined in this set of guidelines. The demands for those services and provisions have tended to increase as communities become more aware of their desirability and availability. Most contracts resulting from this process are beginning to clearly define the cable company's expected performance, but are very vague about the community's commitment and level of participation.

This lack of preparation and commitment on the community's part has two results. If facilities and operational funds are provided but are not extensively used, the cable operators will either close the facilities or reduce their level of operations. After a lengthy introduction period they may rightfully claim they are being forced to provide services no one desires and may ignore the original contract agreement. Once facilities are closed or removed, it is very difficult to have them reopened or restored.

The second result is that the cable operator will use the experience to discourage other communities from requiring similar provisions. Their argument will be "We gave City X all this equipment; they did not use it. Why should we provide it for you here in City Y?"

To avoid this situation, the community must, through a needs assessment, determine the level of potential use of the requested community service provisions. It must determine the minimum level of participation of community agencies and institutions, including allocation of direct funding, staff, and facilities. It may then use that level to establish requirements for channels, equipment, and funding.

Justifying these provisions requires an extensive community education and planning effort, the kind of effort that goes into building a new school, library, or community recreation complex. But the results will be a willingness on the part of the cable operator to provide services and facilities beyond those required. Few operators will hesitate to devote facilities and funding for services that will be available to the community only on their cable system.

Principle #8

The cable company should offer to provide only those community services it is able and willing to deliver.

It is easy for the cable company to provide retransmission of broadcast and satellite services; that is their business. It is more difficult for them to provide community programming and information services because their personnel are not trained in those areas. Yet <u>we</u> see many cable companies promising to provide training for community and institutional access and to provide extensive community programming services. They either make these commitments willingly because they see them as valuable services of half-heartedly because they are required or asked to. Willingly or not, they often assign unskilled and unprepared personnel the task of working with community agencies to produce programs. The result is frustration on both sides. The local school teacher cannot understand why her class must be turned into an elaborate television production or why she cannot borrow a portapak to record a choir performance. The cable company video technician cannot understand why anyone would simply loan an expensive video tape recorder to a person without a background in television production.

The outcome is that equipment does not get used. Programs are not produced. The community is denied access to a communications tool and the cable company loses an excellent public relations opportunity and access to valuable programming.

One solution to this problem that has proved successful is for the cable operator to leave the community programming up to trained individuals from within the community whose mission is to facilitate communications. The cable company role then becomes one of providing funding for the community access centers and distribution systems for their programs.

<u>Principle #9</u>

The franchise contract between the community and the cable company should be opened for periodic review and renegotiation as new communications technologies and community capabilities emerge.

Even the best franchising negotiations will not result in ideal community service provisions and capabilities. Many franchises negotiated in the past 5 to 15 years are woefully inadequate both in terms of the cable system's technical capabilities and community service provisions. Many current franchises are being negotiated and renegotiated with local officials who are unaware of the potential uses of cable television as a communications resource. Few communities have time to develop comprehensive cable utilization strategies before and during the franchising process and frequently discover their own and the cable system's local programming capabilities only after the franchising process is complete.

Since cable companies are understandably unwilling to allocate channels, funding, and facilities to communities that in the past have shown little interest or ability to use the system, those communities frequently find themselves without an access capability. Communities in this position have few options. They can use the company's request for rate increases to open negotiations for increased community services, or they can convince the cable operator that providing support for community programming will result in greater subscriber satisfaction, improved public relations, and a favorable refranchising environment.

Renegotiation in connection with rate increases can result in binding agreements to establish local programming capabilities if the cable company and community agencies and institutions formalize their commitments to provide community access programming. Informal agreements frequently result in sporadic programming efforts and misunderstandings on both sides. Most cable operators are willing to help with local programming efforts if the community's expectations are clearly defined and plans made far enough in advance; but local managers change as do local governments and institutions. For this reason, agreements between the community and the cable operator should be formally adopted in a written form.

II. COMMUNITY ACCESS

Description

"Access" programming as defined earlier is programming produced by community volunteers, nonprofit organizations, and local institutions. The content of the programming is determined by the individual, group, or organization which desires a program and produces that program. There is a wide variety of programming which falls under the umbrella of access, which ranges from public, arts, and minority programming to educational and social services agency programming.

Access Center

In order for access to happen, there must be a place for people to go to learn how to use the tools of television. This is commonly called the community access center or community video center. This facility can be located within the confines of the cable company operation, it can be located under the umbrella of a local institution, or it can be located and operated by a separate non-profit management corporation.

In certain situations across the country, the cable company either voluntarily or through contract requirements in the franchise agreement provides facilities and support mechanisms for access. The access facility is operated out of a building that many times also houses the headend facility or the business office of the cable system. Among the more successful cable company-operated access facilities are those located in East Lansing, Michigan; Encino, California; Atlanta, Georgia; Iowa City, Iowa; Muscatine, Iowa; and San Leandro, California.

The second structure, the institutional alternative, has evolved in communities that have had cable for some time. In these communities, because of interest of a local library, school, or community college, such institution has served as the facility or facilitator for the access programming. We see examples of this kind of structure in operation in communities such as Bloomington, Indiana; Kettering, Ohio; and Rome, Georgia.

The third alternative is the nonprofit management corporation. In some of the older access facilities in the U.S., we find the nonprofit management corporation as the key structure. This approach has evolved from a group of community individuals and/or organizations that felt it necessary to organize a separate corporation to manage the access facility within the community. In many of these communities, this structure has evolved because the cable operator has not been particularly involved or interested in supporting the development of access programming. Some of the communities where we see such structures in place are Reading, Pennsylvania; Madison, Wisconsin; and Marin County, California.

Factors for Success

In order to assure that access programming will develop, your Request for Proposal (RFP) should acknowledge that there are certain factors which are key to the success of access. Those factors include the following:

- o Clear and concise access definition
- o Defined operating structure
- o Specifically designated access channels
- o Appropriate and adequate equipment
- o Appropriate staff
- o Concise, flexible operating rules and procedures
- o Well designed training program
- o Adequate operating budget

The RFP should provide a structured mechanism for the cable company or companies to address these factors. The material which follows can provide such a structure. The form presented is a modification of the forms used by Sacramento, California. CTIC Associates served as the consultant.

Access

A. Access Production Equipment and Facilities

Applicants should list all studio facilities and equipment which will be provided for access production. Please indicate location of access production facilities, and list make, model number, and your approximate cost for each piece of equipment.

1. Access Facilities:
   a. Location (approximate):______________________
   b. Size (approximate sq. ft.):______________________
   Master control: ______________________
   Studio:______________________
   Editing Room:______________________

Access Form, cont'd.

Reception: ______________________________
Maintenance: ______________________________
Other: ______________________________

c. Will this facility also serve as a local origination facility: ( ) yes ( ) no

d. If yes, please indicate specific equipment and facilities available and exclusive of primary access uses.

2. Access Equipment List (provide make, model number, quantity, and your approximate cost for each piece or lot (cables, plugs, accessories). Please list only that equipment which is for exclusive access use.

Note: All equipment (make, model and actual purchase price) listed by the applicant in response to this section must actually be used in construction of the system. If applicant intends to use equivalent items, they must be approved by the city, or its representative, prior to substitution.

B. Access Programming Commitment

1. Total Operating Budget (include only salaries, benefits, maintenance of production equipment, tape stock, miscellaneous supplies and access promotion):

Year		Year		Year	
1	$____	6	$____	11	$____
2	$____	7	$____	12	$____
3	$____	8	$____	13	$____
4	$____	9	$____	14	$____
5	$____	10	$____	15	$____

2. Staff Commitment:

	Full-time personnel	Part-time personnel
a. First year		
b. Third year		
c. Fifth year		
d. Tenth year		

Attach a narrative statement describing access staff job assignments.

Access Form - Cont'd

3. Discuss local programming philosophy specifically describing plans and objectives for access, community use, and local origination. Include the names of organizations and individuals contacted by the applicant to assess local needs and desires.

C. Access Channels and Administration

1. Applicant is to attach a complete set of rules and procedures for the operation of public, educational, government, and leased access channels. The rules must describe the following:

- o availability of channels to various users;
- o availability of equipment and rules governing use of equipment;
- o scheduling procedures for reserving equipment and channel time;
- o any rates to be charged, including deposits;
- o copies of contract forms and application forms; and
- o availability of production assistance, etc.

2. State the nature and extent of all training to be offered by the Franchise respecting equipment operation and training required as a condition of facility and equipment use and operation by users.

3. Describe the independent body (if any) proposed by the applicant to administer access programming.

a. Legal form of existence;

b. How established and who will be responsible for establishment;

c. The size, composition, and method of selection of the Board of Directors;

d. The terms of Board members, and grounds and procedures for removal of members, if any;

e. The specific powers of the body in relation to administration of community use and the means by which such powers may be exercised and enforced;

f. Sources and amounts of funding for support of operation of the body.

Access Form - Cont'd

4. Describe any standards or criteria which you intend to utilize in connection with the following issues:

a. The time made available for, and community use programming covering candidates for public elective office during election campaigns;

b. Program quality control;

c. The legality of program content and violation of the legal rights of others;

d. Any and all pre-conditions of whatever kind or nature relating to use by third parties of studio facilities or production equipment and broadcast of programming presented thereby.

e. Determinations relating to the tier of service in which community use programming by local governmental agencies and local nonprofit community organizations will be placed.

III. LOCAL ORIGINATION

Description

Local origination programming, as mentioned earlier, is programming which is locally produced or locally originated (i.e., a purchased program which is "played-back" from the cable company's control room). The program content is determined by the cable company and local production is undertaken by cable company employees. The channel is commercial and thus local businesses are approached to purchase spot advertising.

A new aspect of the sales and marketing area of local origination includes the sale of the "advertising availabilities" provided by a number of satellite services to the local cable operator. Through innovative packaging of the local origination programs and satellite services, the cable operator can provide local businesses with the opportunity (at affordable prices) to advertise on local origination programs and on satellite delivered programming.

If you desire local origination channels on your cable system, the following form will be helpful in gaining information from the cable company or companies in an organized and structured manner.

LOCAL ORIGINATION PRODUCTION EQUIPMENT AND FACILITIES

Applicants should list all studio facilities and equipment which will be provided for local origination production. Although facilities may be available to all classes of users, please note that access studios and equipment are for exclusive use of access channel users, and should exclude any facilities or equipment listed on this page.

Please indicate location of production facilities and list make, model, number, and your appropriate cost for each piece or loss of equipment.

1. Local Origination Facilities:

a. Location (approximate):______________________________

____

b. Size (approximate sq. ft.):________________________

____

Master Control:______________________________

Studio:_____________________________________

Editing Room:________________________________

Reception:___________________________________

Maintenance:_________________________________

Other:______________________________________

c. Will this facility also serve as an access studio?

( ) yes ( ) no

d. If yes, please indicate specific equipment and facilities available and access uses or use of LO facilities by access used and attach copy of any application form.

2. Local Origination Equipment List (provide make, model number, quantity, and your approximate cost for each piece or lot (cables, plugs, accessories.) Do not include equipment available to access channel users on an exclusive basis.

Note: All equipment (make, model and actual purchase price) listed by the applicant in response to this section must actually be used in construction of the system. If applicant intends to use equivalent items, they must be approved by the city, or its representative, prior to substitution.

IV. MUNICIPAL/GOVERNMENTAL ACCESS

Description

A cable communications system can be one of the main mechanisms used within a city to deliver total community services, information, and data. Since local government is frequently the primary provider of services to many communities, it is imperative that municipal staff learn how to effectively use cable technology to assist in delivery of services and municipal information. Further, local government officials should be aware of the impact of any decision they make regarding cable communications systems in the city.

One of the oldest municipal access operations in the country is in Madison, Wisconsin, in operation since 1974. City 12 provides cable service 24 hours a day. Programming icludes City Council meetings, Board and Commission meetings, annual budget hearings, city events, local election returns, alphanumeric announcements, and information.

The Miami Valley Cable Television Council, a six city cooperative project, provides training facilitation services to suburbs south of Dayton, Ohio. The services include 24 hours a day of programming, which includes the following: live City Council meetings from five of the municipalities, weekly information programs produced by different city departments, such as recreation, planning, city manager, police, fire, etc; seasonal specials on topics such as Halloween safety tips for kids from the police departments or spring clean-up information from the public works department; regional hearings on topics of area concern.

During the franchising or renegotiation process, cities should access their communications resources and needs. City departments should be asked to develop position papers which state the manner in which the department wishes to use the cable communications system. These papers should not be extensive shopping lists, but rather indications of departmental daily functionings and how these can be more effectively or cost-efficiently delivered by a cable communication system. Further, the position papers should discuss areas of potential future services the department hopes to explore through cable communications.

A municipality should set up guidelines or operating rules and procedures which can provide basic direction for the municipal access channel. These guidelines should include at least the following items:

(1) Objectives; (2) Overall policy; (3) Operational procedures; (4) Access policy; and (5) Editing policy. On the following pages you will find examples of operating rules and procedures.

Municipal Access Operating Rules and Procedures

1. Objectives

The primary objectives of the Municipal Cable Channel should include:

A. To provide public service information to the citizens;

B. To widen the dissemination of information regarding the activities of the legislative and advisory bodies;

C. To increase the knowledge of the citizens as to the various functions performed by their city government;

D. To provide additional information to citizens needing access to the various city departments;

E. To assist in internal training of appropriate city departments;

F. To assist in internal information distribution.

2. Overall Policy

Programming - shall provide direct, non-editorial information to the citizens concerning the operations and deliberations of their city government.

3. Operational Procedures

A. Modes of Operation

1) Live Cablecast - This may consist of cablecasts of the City Council meetings and other public meetings and events of general community interest at the discretion of the community.

2) Tape Delayed Cablecast - Many public meetings and events will be videotaped for cablecasting at a later date. Some meetings such as City Council, will be cablecast both live and subsequently by tape at other convenient times during the week.

3) Locally Produced Programs - Programs will be produced locally to illustrate the functions or operations of some form of City Government. These might include videotape tours of government facilities such as parks, libraries, or the sewage treatment plant, or might be on a specific city program such as streets, or building inspection.

4) Outside Source Programs - Much material concerning local government operations is available elsewhere in the country and may be borrowed for local use. This material will be used when appropriate.

5) Character Generated Announcements - During all hours of operation when no other programming is scheduled, a character generator will provide a continuous display of current messages of interest to the public.

6) Internal Training and Information Exchange - Many internal city usages of cable are possible such as:

a) Training for city personnel;

b) Information exchange on matters such as personnel programs, parks, and recreation scheduling,

c) Information exchange between local police and fire departments.

B. Access Policy

1) Public Meetings - All public meetings of city policy-making or advisory boards and commissions are authorized for cablecasting. All taping or live cablecasting of such meetings will be coordinated in advance with the board chairperson and the City Council.

2) Information Programming - All city departments may submit requests for programming which they feel appropriate for the municipal channel. These may be locally produced or may be obtained from outside sources. Only those tapes which are consistent with the overall operating policy of the cable channel shall be cablecast.

3) Character Generated Announcements - Information for the character generated announcement may be submitted by any city department. Announcements should be in keeping with the intent of this policy statement.

C. Editing Policy

1) Public Meetings - Any public meetings cablecast shall not be edited or subjected to editorial comment. Meeting coverage shall be from gavel to gavel.

2) Departmental Programs - Any programming prepared by or provided by an individual city department may be modified or edited as appropriate and as dictated by scheduling and manpower availability.

3) Character Generated Messages - Announcements programmed in the character generator shall be edited to provide clarity and to maximize the use of the character generator.

4) Error - Should human error result in the cablecast of incorrect information over the cable channel, the city and the employees thereof, shall not be liable for the inaccuracy of the information.

D. Endorsements - At no time will the channel endorse specific brand names of products for consumer use.

E. Use of Outside Resources - In order to maximize programming, every attempt will be made to use outside community resources to assist in channel utilization, for instance, using student interns and work-study students to assist in special projects as well as to assist in production cablecast of City Council and other meetings.

F. Channel Operating Hours - It shall be the general goal of the municipal channels to have some form of programming available continuously. The general approach will be to utilize live and taped programming when available during weekdays, and to have a continuous character-generated announcement service at all other hours, 24 hours a day.

G. Retention of Tapes - It shall be the general policy to retain video tapes of locally produced events and meetings for a 3-week period. At the end of that time, the tapes may be reused and the original material erased. Any requests for longer retention should be made in advance of the 3-week limit.

V. INSTITUTIONAL NETWORKS/ORIGINATION POINTS

Description

An "institutional network" is specific channel/spectrum space separate from, but capable of interconnection to, the home subscriber services. This network is used to carry video, audio, and data signals to or from nonprofit or commercial institutions in the service area. For example, the institutional network might be used by schools or hospitals to offer training seminars to professional staff members. City governments could use the network data capabilities to link computers in several public buildings. Community groups might use the network's origination points to provide a live feed of a special activity such as a city festival to viewers on the home subscriber network.

There are two basic technical approaches that cable companies will use to offer community institutional networks. Under both techniques companies can divide channel spectrum into full 6MHz video/audio and 3KHz voice/data facsimile channels.

1. Integrated Networks: It is possible in small systems using a single trunk to assign institutional functions to spectrum space on the single subscriber loop. This alternative, while being better than having no service, will allow extremely limited use of the network, especially in the areas of teleconferencing and multiple origination. (For instance, for a live teleconference from four locations about five full 6 MHz channels will be needed on the institutional allocation.)

2. "Dedicated Network": In most new cable proposals, companies are offering to install a totally separate cable, connecting all institutions. By having a separate but interconnected institutional cable, communities can have anywhere from 30 to 90 channels for institutional use.

The allocation of channels on these networks will be based on the requirements of the institutions and the cost of leasing space on the available spectrums. Before a local institutional network can be designed, it is critical that a local community needs assessment be carried out to determine specific projected utilization. Local institutions may decide to use the network to:

A. Interconnect hospitals and health care centers with medical education and diagnostic facilities;

B. Provide closed circuit counselling sessions for mental health centers;

C. Provide in-service training for police and fire departments;

D. Provide telemetry services to monitor water, sewer, energy, and traffic flows;

E. Deliver educational materials to classrooms in public schools, vocational schools, colleges, and universities;

F. Transmit patient and client records among health care institutions and social service agencies;

G. Provide teleconferencing from all institutional centers;

H. Provide live coverage of community events from multiple origination points;

I. Provide institutional connections with satellite uplink/downlink facilities;

J. Provide library catalog and reference services;

K. Transfer funds among financial institutions;

L. Offer a variety of business communications services such as transmitting inventory information from retail outlet to home office and warehouse.

Since this list provides minimal suggested utilization, it becomes clear that system design and capacity will have to be determined after a local needs assessment.

Municipalities should be aware that institutional networks, while providing significant services, involve a significant capital investment with on-going maintenance costs. It is critical that when cable companies submit proposals, they specify the exact cost for construction and operation of the institutional network, along with specific plans, letters demonstrating commitment or strong interest from local institutions, and marketing plans that will generate revenue to pay for the network.

It is important to differentiate between use of the institutional network by profit (commercial) vs. nonprofit (non-commercial) institutions. Three critical issues which must be addressed in any company proposal are:

A. Will nonprofit organizations be guaranteed access to the network channels?

B. What specific rate reductions, waivers of fees, and/or equipment is being provided to nonprofit organizations?

C. Will home subscriber revenues be used to subsidize (on a short or long term basis) a network that will be providing low cost services to profit making institutions?

In regard to question #1, it is critical that municipalities provide protection for nonprofit access to the institutional network. As cable companies begin to lease channel spectrum to banks, insurance companies, retail stores, and other businesses, city councils may find that the number of nonprofit access channels are reduced to allow expansion of revenue-generating commercial channels. As a general rule, communities should try to negotiate for at least a 50/50 split between profit and nonprofit access.

Question #2 raises two issues: having minimal necessary equipment provided and "rent an institution." It is critical that cable company proposals include all necessary modulators, scrambling devices, converters, etc. that institutions will need to use the network. There have been many cases where institutions have been ready to go "on line" only to discover that a $4,000 or $20,000 item has not been provided. A second concern here is a phenomenon called "rent an institution." Sometimes the cable company will offer, and in many cases local institutions will demand, thousands of dollars in equipment to secure the franchise. Municipalities need to be aware that special considerations for any institution will be paid for by all parties to the cable system, creating, in effect, a subsidy for that institution. As long as the community is aware of the cost, and a justifying benefit can be identified, the community should feel free to treat the subsidy as it would any other support for local public services.

To permit maximum flexibility in the institutional network and to ensure the right of non-commercial and public service institutions and organizations to that network, the NFLCP recommends the following:

- That the network be dedicated rather than integrated
- For large cities where the needs assessment has indicated active use of the institutional network, that the network be reserved fully for public service non-commercial use
- For municipalities where needs may not require the full network, that a portion not less than 50 percent be reserved for public service non-commercial use
- That nonprofit organizations and non-commercial institutions be guaranteed access to the network
- That the <u>community</u> (usually the community access organization) reserve the right to establish criteria and priorities for access to the non-commercial network should it become crowded
- That full switching capability from the institutional network to the subscriber network be required
- That an adequate trial period for the network be established which will include:

-- Promotions and community education provided by both the cable operator and the community
-- Use of the network by non-commercial users at low or no cost
-- A moratorium on reassigning any space that is not used by the institution originally assigned that space

o That any public service non-commercial institution that loses its allocated space due to non-use be allowed to regain that space when it can demonstrate readiness to use the space

o That expansion of the network in the future be planned for, including:

-- the specific technical procedures to be followed to create expansion
-- the costs for establishing additional channels (drops, transmission and switching equipment, etc.)
-- who will pay the costs

o That the community reserve the right to require expansion of the network if it is justifiable

o That there be adequate space reserved for non-commercial use in any planned expansion

o That nonprofit community service organizations be guaranteed preferred rates for leasing spectrum space (possibly $1/year)

o That nonprofit community service organizations be able to obtain additional services for free, at nominal rates, or at preferred rates, not to exceed the actual cost to the cable operator for providing the services, including staff needed to provide the services or coordinate use of the network

o That (if permissible by state and federal regulation) any revenues guaranteed by use of the institutional network be included in the revenue base for the franchise fee

o That no non-commercial institution or organization in the community be granted special privileges to use the network or to be equipped or staffed to use the network unless those special privileges are justified by public services to be provided by that institution's use of the network

o That commercial users of the institutional network be given fair opportunity under consistent criteria to gain access to the network

o That the municipality require channel capacity on the institutional network to be increased when it is clear that the capacity of the network is being taxed. It may be necessary for the authority that assigns channels on the institutional network to first re-assign some channel time from light users to heavy users.

Recommended RFP Language:

Institutional Network/Origination Points

General Description:

The city is interested in having an institutional network, separate from but interconnected to the home subscriber service, to provide video, audio, and data services to and from local institutions. Applicants should propose an institutional system that will meet criteria established by local community needs assessment.

The following represents a list of institutions which may be considered as part of an institutional network. It is not a commprehensive list and applicants may add or delete as cable needs are ascertained.

Nonprofit (Non-commercial) Institutions:

- Universities
- Private and commuity colleges
- Public, private, and parochial schools
- Arts and culture centers
- Community centers
- Nonprofit corporations desiring to develop teleconferencing and/or linking computers
- Video centers
- Libraries
- Hospitals, community health clinics, and other nonprofit health care facilities
- Parks and recreation buildings
- Private nonprofit institutions (YMCA and Salvation Army, etc.)
- Government agencies such as Agricultural extension offices, regional educational facilities, post offices, senior citizens centers, humane societies, etc.
- Police and fire departments
- Religious institutions
- Natural community gathering places

Profit (Commercial) Institutions:

- Financial institutions
- Insurance companies
- Retail stores
- Medical facilities
- Transportation facilities
- Hotels/motels
- Other private businesses

RFP Questions:

1. Will the cable company provide "institutional network" services? Yes ____ No ____ If yes, continue. If no, justify through documented community ascertainment studies, specific financial pro forma information why it is not technically or financially feasible to offer these services.

Note: All equipment (make, model and actual purchase price) listed by the applicant in response to this section must actually be used on construction of the system. If applicant intends to use equivalent items, they must be approved by the city, or its representative, prior to substitution.

System Design

2. Describe in detail the proposed institutional network design. Clearly indicate whether institutional services will be provided through utilization of spectrum space on the subscriber loop or if a separate cable or cables will be used. Include all equipment, hardware, and materials to be used in the entire institutional network with notations for make, model, description, and price.

3. Information provided should include, but not be limited to, detailed maps indicating locations of cables, amplifiers, hubs, etc. The description should also include a projected service plan and uses of the institutional network (e.g., teleconferencing, video programming, high/low speed data, security services, etc.) This section should include specific technical description on how each proposed service will be carried out. Use the form below to describe channel allocation and capacity.

In addition to the previous institutional network design information, provide responses to the following:

(Note: List specific equipment, including make, model description and price for all sections. Do not simply refer to portions of overall system description.)

4. a) Describe in detail addressable taps or other centrally controllable devices that will be provided initially to offer discrete channel distribution on either the A trunk (integrated network) or the B trunk (dedicated network).

b) Describe in detail headend scrambling equipment and converters that will be available or provided to institutional users who require confidential distribution of programs, teleconferences, etc.

5. Describe in detail how and which equipment will be used to provide for transmission and reception of video, audio, and data signals

INSTITUTIONAL NETWORK-CHANNEL ALLOCATION

	Example
Cable Channel	A
MHz Range	30-36 MHz
Use: Commercial Non-commercial Reserved for Future Use Down Stream	Commercial
Up or Down Stream	Upstream
Specific or planned institution by service category	National Bank of
Mode Video/Audio bank	Data
Activation Date and duration	6 mos. after basic service-entire length of franchise

via appropriate upstream/downstream frequency conversions and channel patching at the headend, the hubs, or both.

6. Describe in detail all switching equipment that will be provided to patch channels from the institutional network to the subscriber network and vice versa.

Channel Assignment

7. Outline specific cable company policy governing channel assignment. In responding to sections a, b, and c, offer specific language rather than a general "when need is demonstrated" provision.

a. Under what conditions would additional channels be granted?

b. Under what conditions would a channel assignment be rescinded?

c. Under what conditions would a channel assignment be transferred (not only in relation to spectrum space, but also in relation to tier placement)?

d. Specifically, which parties will have responsibility, accountability, and final authority over channel assignments?

Channel Access

8. Outline channel access policies:

a. Describe in detail how priorities for channel use will be determined for the institutional network.

b) Specify provisions regarding guaranteed access to the institutional network for nonprofit users.

c) Specify provisions establishing criteria for access for profit (commercial) users.

9. Will any channels on the institutional network be reserved for use by the cable operator or other commercial uses? Explain in detail.

10. List the institutions proposed to be included in the institutional network. Clearly indicate whether the institutions listed will be initially connected or simply passed by cable.

11. Give details of any existing or proposed agreements with or commitments to any institutional network user(s). Attach any such letters or agreements as appendices to this application.

12. Using the table below, describe in detail all equipment that will be provided to local institutions for video, audio, and data.

EQUIPMENT TO BE PROVIDED FOR LOCAL INSTITUTIONS:

Institution	Equipment	Quantity	Model	Description	Cost	Purchased by	Maintained by

Funding

13. Describe in detail projected charges for users on the institutional network. Itemize all services and related changes. Response should include but not be limited to the following:

a) Will there be a charge to all leased access users?

b) Will nonprofit users be offered reduced rates or free services?

c) Will a portion of the home subscriber revenues be used to pay for institutional network construction/operation?

14. Will there be an initial demonstration period in which no charges will be applied for use of the institutional network? If yes, for how long and under what conditions?

15. Is it projected that the institutional network will become self sustaining? Complete in detail the following financial pro forma (Institutional Network). Include all projected related expenses and revenues for construction and operating including, but not limited to, indicated categories.

Construction and Activation Schedule

16. Provide detailed construction and activation schedules for every component and service described in your proposal for an institutional network.

Recommended Franchise Language:

Institutional Networks/Origination Points

The company will provide an "institutional network" serving

INSTITUTIONAL NETWORK:

Annual Budget Expenses	Year 1	2	3	4	5	6	7	8	9	10	11-15
Salaries											
Benefits											
Equipment (Specify)											
Rent											
Maintenance											
Marketing/Promotion											
Training & Education											
Supplies											
Grants/Awards											
Travel											
Other-Specify											
Capital Expenses											
Modulators											
Switching Equipment											
Interconnection Equipment											
Central Control Equipment											
Cable/Installation											
Other-specify											
Income											
Channel Costs											
Equipment Lease/Rental											
Switching Services											
Interconnection Charges											
On-line Computer Lease Costs											
Other-specify											

public and private institutions and businesses. The institutional network shall be capable of transmission and reception of digital, audio, and video signals among any or all connected institutions via appropriate upstream/downstream frequency conversions and channel switching at the headend, the hubs or both. This shall also be the capability of switching channels from the institutional network to the subscriber network and vice-versa.

(Optional language 1--for very small communities)

The institutional network shall consist of a minimum of ______[1] upstream and ______[1] downstream channels utilizing an integrated system on the home subscriber network ("A" trunk) as described in the company's proposal. A minimum of 50 percent of available institutional channels/spectrum will be allocated on a priority basis to nonprofit institutional users.

(Optional language 2--for medium to large communities)

The institutional network shall consist of a minimum of ______[2] upstream and ______[2] downstream channels on a second, dedicated cable or cables, separate from but interconnected to the subscriber network, as described in the company's proposal. A minimum of 50 percent of available institutional channels/spectrum will be reserved for nonprofit use.

The company shall provide addressable taps or other centrally controllable devices and/or scrambling equipment and converters, as described in the company proposal that will allow discrete channel distribution on both the institutional and subscriber networks.

The company shall provide a written policy, approved by the city cable commission, which governs channel access and assignment on the institutional network, including but not limited to provisions for protection of nonprofit user access. The company shall add one channel to the institutional network whenever all institutional channels assigned to nonprofit users are in use during 60 percent of the time between the hours of 9:00 a.m. to 6:00 p.m. and 80 percent of the time between the hours of 6:00 p.m. and 9:00 p.m. on weekdays (Monday through Friday) up to the channel capacity of the system.

The company will be required, at its sole expense and without charge, to provide a service drop with converter and ongoing basic services to each public and/or nonprofit institution designated by the cable commission at the time these facilities are passed by the transmission cable. These drops shall include both the institutional network channels and all home subscriber "basic service"[3] channels. The franchise is also required, upon request, to provide consultation services and advice on internal communications systems and to sell equipment and service and provide installation of the systems at charges not to exceed actual out of pocket costs.

The institutional network shall be subject to the same requirements for interconnection as specified in section ____ of this ordinance.[4]

All requirements including but not limited to institutional network construction, activation dates, equipment to be provided, etc., as required by the ordinance and stated in the company proposal shall be strictly adhered to. Failure to comply with requirements will subject company to the enforcement provisions stated in section ____[4] of the ordinance.

---

1. For very small communities the minimum would be at least one upstream and one downstream channel. Two upstream and two downstream is a preferred minimum to allow equal access for both commercial and nonprofit use.
2. These minimums will vary depending on total institutional network capacity. Dedicated network channel capacity will range from 12 to 26 channels upstream and 11 to 39 channels downstream.
3. If the company is offering "tiered" services, you should specify here which tier you feel is most appropriate to be supplied, without charge, to local institutions.
4. See "interconnection."

VI. TWO-WAY SERVICES

Description:

The term "two way services" as used here means any service which is dependent on the simultaneous origination and reception of signals regardless of whether or not such simultaneous signals travel on the same bi-directional cable or are switched from different cable sources at a central point. The category of two-way services is conceived here to consist of two types:

1. Security/consumer services, of a response or feedback nature and primarily narrowband. These services may be either automated or non-automated.

A. Automated-burglar, fire and status monitoring systems which operate on the principle of automatic, periodic interrogation of a home terminal by a central computer.

B. Non automated services which require action on the part of a subscriber in order to input a signal to the central computer (e.g., medical alert, polling, funds transfer, data retrieval, etc).

2. Interactive services that permit the combination of video and audio from any two of multiple locations on a single downstream channel for display on a receiver's screen. This enables two parties engaged in an interactive process (speaker and listener) to see each other face-to-face "through" their respective television receivers. This type of interactive mode may be thought of as "two-way audio/video teleconferencing."

Responsive/feedback services:

Security/consumer services can best be termed feedback or response systems. Such systems utilize a computer at the cable company headend and a terminal in the subscriber's home or place of business. The computer is programmed to "sweep" the subscriber system periodically, interrogating each terminal. The terminal is programmed to respond to the computer if certain pre-determined parameters are met.

These parameters are determined by the level of service the subscriber has purchased. For instance, a terminal response to the computer may be triggered by the presence of smoke detected by one or more of a series of sensors, of the opening of a door or window detected by an intrusion sensor. A third security service is medical alert. In this case, a switch or button is placed in some easily accessible location in the subscriber's home which when activated, will summon medical assistance. Other, non-automated, two way services include polling, shopping by cable, funds transfer, data retrieval, etc. In these cases, a key pad, similar in appearance to a hand-held calculator, is used by the subscriber to respond to questions, register an order for merchandise, transfer funds from one account to another, order text information for display on the TV set or order pay-per-view programming, by punching pre-determined combinations of numbers. Computer-aided instruction, electronic mail, meter reading, and traffic light control are also two-way services that have been proposed.

It should be noted that computer-based two-way services are highly capital intensive. Therefore, their use for nonprofit public interest purposes is less likely to develop than their use as a revenue generating tool. However, an NSF experiment in computer-assisted learning was conducted in which fire training was provided to members of the Rockford, Illinois, fire department.

While the possibility of these services is impressive and commercially attractive, many of them have been tested only on a small-scale experimental basis and others have not been tested at all. Before development of the RFP, the city should research the status of these services. Computer based response/feedback technology is expensive. Installation fees for security services may exceed $1,000 for an average home in addition to monthly service charges of $12 or more. Polling ability is usually placed on the most expensive level of basic cable service.

The future development of these services in a particular city poses serious and far-reaching questions which must be carefully and fully examined during the franchising process. The societal effects of these services have not yet been assessed let alone experienced. For instance, what are the implications for retail establishments in an urban area if shopping from home becomes a significant means of fulfilling consumer needs? In regard to polling, issues such as privacy and selective citizen participation arise. What are the implications inherent in a computer which can store information on how a household voted on a controversial question? What are the implications of cities' possibly basing decisions on the opinions of a segment of the community which is pre-selected due to its ability to pay for this kind of participation through a costly cable service?

These questions will be addressed below in sample RFP and ordinance language. However, they must be addressed in detail in each community.

Interactive Services

Interactive services, while a form of two-way cable, differ radically from those described above. An interactive system transmits two-way audio and video, rather than data, in such a way that a speaker and listener at widely separated locations can interact face-to-face. An interactive network can consist of any number of locations. Switching equipment at the headend permits any combination of two sites to be connected by two-way video and audio. This combination can be changed instantaneously to follow the flow of interaction among the sites. Examples of some uses to which this interactive system has been put are:

1. Senior citizens in hi-rise apartments can have face-to-face contact with the Social Security director in his office from the community room of their building.

2. City government can conduct public hearings using several locations around the city tied together with interactive cable. Residents can attend the hearing and participate directly by going to a neighborhood center near their home. Cable subscribers can participate from home by telephone.

3. Institutions can use the interactive network to share in-service training with one resource person simultaneously training the staffs at several institutions.

4. Continuing education instructors can teach classes at several locations at the same time.

5. Through the use of neighborhood origination centers, individuals who are not cable subscribers can participate in the interactive process.

These are only a few examples of the use of interactive technology. Unlike the response systems described above, interactive services require relatively inexpensive equipment and operators can be easily trained. In Reading, Pennsylvania, 1300 hours of this programming is done annually by a nonprofit corporation supported totally by local funds and a contribution from the cable company. The heavy use of this existing interactive system represents considerable savings in transportation electronically from near their homes or places of business to interact with others up to 20 miles away.

There are several significant differences between these two types of two-way services. Security/consumer services, including polling, are normally only available on the highest level of basic services, or are offered only in addition to basic service. Interactive programming can be available on the basic level as a portion of the community access channels. The ease of establishing neighborhood interactive sites makes this programming accessible to those who are not cable subscribers. Therefore, neither low-income individuals nor individuals who are not subscribers are excluded from the interactive process. Interactive services, as a result, tend to promote community dialogue and cohesiveness and information exchange while polling type services tend to further separate the "information rich" from the "information poor" and place an economic barrier on cable mediated participation in community processes.

While both services appear, on the surface, to permit an increase in mobility through cable and a breakdown of isolating barriers, closer examination suggests that this is not necessarily the case. Security/consumer services may tend to increase isolation from society by encouraging a "fortress mentality" among that segment of the population which is able to afford them. While polling services appear to encourage greater participation in the public process, the fact that the subscriber can only respond to a question formulated by another in a "yes-no" or "for-against" manner leads to the oversimplification of complex issues. For example, "Should a freeway be built through the city - yes or no" "Should capital punishment be mandatory for murder-yes or no." What is missing here is what forms the heart of interactive services. That is open dialogue among members of the community each of whom has the ability to articulate his or her own comments, qualify his or her own answers, and communicate as an individual to other individuals rather than responding as an undifferentiated member of a collective group.

Research conducted in Reading, Pennsylvania, by New York University indicates that interactive services do in fact decrease the isolation of the elderly, homebound, etc., by enabling them to interact with others face-to-face as individuals. The type of two-way service, or combination of types, selected by a city, will have a real effect on the quality of life of its citizens over the term of the franchise and must be carefully and seriously considered.

Recommended RFP Language:

Two-Way Services

General Description

The city is interested in having two-way services that allow both the transmission of data from the home/institution to a central computer and the transmission of audio and video signals in such a way that a speaker and listener at widely separated locations can interact face-to-face. Applicants should propose two-way services that will meet criteria established by local community needs assessments.

RFP Questions

1. What two way security/consumer services will be provided?
   a. intrusion alarm
   b. fire alarm
   c. medical alert
   d. funds transfer
   e. teletext
   f. viewdata
   g. shopping
   h. polling
   i. computer assisted learning
   j. meter reading
   k. energy monitoring
   l. traffic control
   m. other ____________________

2. What equipment will be used to implement the above services? Complete the chart that follows.

Chart for Question #2. What equipment will be used to implement the above services

Equipment Item	Quantity	Manufacturer	Model	Purchase Price	Date of Availability	Type of Service

3. For each service proposed, list the date of activation and duration.

Service	Area of City	Activation Data	Duration

4. Describe in detail what safeguards will be instituted in the event of a system outage of security services.

5. Describe in detail the duties and responsibilities of security center staff. Will the security center be staffed 24 hours per day, seven days per week?

6. Describe in detail what safeguards will be instituted to guard against false security alarms.

7. Describe in detail the market studies which indicated the feasibility of the above proposed services.

8. Describe in detail the procedures which will be instituted to ensure the privacy of computer data.

9. If polling service is proposed, will equipment be provided to access centers and neighborhood centers to permit residents who don't subscribe to cable services to participate in public hearings? Describe in detail and include a map showing the location of each site and the equipment to be provided.

10. What equipment will be provided to nonprofit institutions to enable them to use the above proposed services?

11. Will interactive two-way audio/video capability be provided?

12. If so, describe in detail what headend and other equipment will be used to implement the services on the chart that follows.

Chart for Question #12. Describe in detail what headend and other equipment will be used to implement interactive two-way audio/video services

Equipment Item	Quantity	Manufacturer	Model	Purchase Price	Date of Availability	Location

13. Will this equipment be made part of the access package for use by the access corporation? If no, describe in detail the percent of time this equipment will be dedicated to nonprofit use and the procedures used to ensure equitable sharing.

14. Will a separate, dedicated return system be used?

15. If no, how many channels will be allocated for interactive return purposes?

	Commercial Channel/Frequency	Non-profit Channel/Frequency
Subscriber loop		
1. Video/audio		
2. Audio		
3. Data		
Institutional loop		
1. Video/audio		
2. Audio		
3. Data		

16. What locations will be connected for participation in interactive programming? Include a map with locations clearly indicated.

Sites	Date of Activation	Return Channels

17. What will be the total number of return channels available for simultaneous origination? Include maps.

A. To hubs	Channel	Frequency	Location
Hub #1			
Hub #2			
Hub #3			
etc.			

B. From hubs to hubend	Channel	Frequency	Location
Hub #1			
Hub #2			
Hub #3			
etc.			

18. Describe in detail how the switching of interactive programming will be coordinated.

SUGGESTED ORDINANCE LANGUAGE - TWO-WAY SERVICES

Security/Consumer Services

1. During the operation of the cable communications system, the franchisee shall strictly observe the privacy and property rights of subscribers.

A. The following restrictions shall apply to the release by the franchisee of information and data:

1) The franchisee shall always indicate the number of its subscribers in the City.

2) When indicating the number of its subscribers viewing a particular channel at a particular time, the franchisee shall indicate the total number of subscribers viewing during the relevant time and the

percentage of all subscribers which they represent. In no event shall the franchisee release the viewing habits or preferences of a particular subscriber.

3) Any poll conducted by the franchisee to determine subscriber preferences shall indicate by a whole number those subscribers expressing a particular preference and shall never be expressed as a percentage of subscribers expressing that preference.

4) The franchisee shall at all times release both the number and the percentage of subscribers purchasing any service contemplated in this chapter, but not the indentity of any subscriber.

5) The franchisee may maintain such records as are necessary to bill subscribers for the purchase of any cable communications service.

6) Neither the franchisee nor any other person shall initiate in any form the discovery of any information on or about a subscriber's premises without prior written authorization from the subscriber potentially affected.

7) Valid authorization shall mean voluntary written approval and consent obtained from a subscriber to conduct the investigation described in subsection 1A(6) hereof, prior to the commencement of such investigation for a period of time not to exceed one (1) year from the date of such authorization, and that shall not have been obtained as a condition of cable communications service or continuation thereof or communications facility usage.

B. Without the authorization described in subsection 1A(6) hereof, neither the franchisee nor any other person shall in any manner activate, utilize or otherwise operate any channel from a subscriber's location.

C. Every subscriber shall have the absolute right to deactivate the return path from the subscriber's receiver at the franchisee's sole cost.

D. The franchisee shall not tabulate any test results, not permit the use of its cable communications system for such tabulation, which would reveal the commercial product preferences or opinions of subscribers, members of their families or their invitees, licensees or employees.

E. Violations of any provision of this section shall be considered a material breach of any agreement awarding a franchisee in accordance herewith and shall subject the franchisee to the provisions of section 425.25(e) and to all penalties and remedies prescribed in such agreement as well as all other legal or equitable remedies available to the City.

2. In the event of a system outage of any security service the franchisee will immediately notify the appropriate authority (fire, police, hospital) regarding the exact geographic area of the outage and estimated repair time.

3. The franchisee will staff the security systems monitoring center on a full-time basis (24 hours per day, seven days per week). Such staff will have no other headend duties.

4. Accurate time logs of the activities of security monitoring personnel will be maintained for review by regulatory authorities.

Interactive Services

1. The franchisee will equip the headend and hub sites to permit the demodulation, monitoring, and switching of interactive programs utilizing split screens both on a segmented hub basis and system wide. Priority use of this equipment will be made available to the community access corporation.

2. Audio/video transmission capability (exclusive of origination equipment) will be provided from the following:

A. All government buildings;
B. All libraries and branches;
C. All access centers;
D. All school buildings;
E. (Other neighborhood centers should be included determined by a community ascertainment study.)

3. Portable, live origination units will be provided to the access corporation to enable interactive participation when acquired from those locations specified in #2 above. These units must be capable of being transported in a private car. (The number of units should be determined by a community ascertainment study.)

4. Return channel capacity will be reserved for use by nonprofit institutions and access centers (at a level commensurate with the result of ascertainment study).

5. The franchisee will provide a secure means of transmission to enable those nonprofit institutions requiring confidentiality to utilize the two-way audio/video mode with security (e.g., mental health, medical, legal).

VII. AUDIO SERVICES

Description

When referring to cable, most people think only of video services. Cable also provides the opportunity for delivering high quality audio signals which may be heard on connected standard FM receivers, as background audio on television channels, or on special receivers. Among the varied sources of programming are local and distant stations, satellite delivered signals, and community produced programs on leased or cable company-provided access channels.

Several problems exist in providing audio services:

1. While radio is an older medium than television, the larger number of local radio stations has inhibited the development of nationally distributed audio services and model community access services using audio. This situation may change.

2. Some subscribers to cable may be more interested in audio than video services. For audiophiles a crisp stereo signal is essential. These same individuals have pet stations and formats which can be imported if not available locally. Include questions in the needs assessment to determine that, where technically possible, these stations can be included.

3. The availability of at least a single audio community programming channel is suggested. This will require sufficient equipment to make such a service viable. The type and amount of equipment depends on the service desired. Recording programs, remote sports events, city council meetings, and call-in programs all require different complements of equipment. A visit to your local radio stations for information may be a guide. In fact, your local commercial or public radio stations may have underutilized production capacity which might be leased for this purpose.

4. Cable subscribers can only pick up sub-carrier signals if the carrier-to-noise ratios, spurious signal levels and injection levels are proper. Every FM station has the benefit of the subscriber to ensure that access to all audio services is possible.

Recommended RFP Language

Audio Signal Carriage and Channel Allocations

1. On the following chart, provide the requested information for each audio service channel on the proposed system. Complete a separate

chart for each type listed below. Reproduce chart as needed.

A. Local FM Broadcast
B. Imported FM Broadcast
C. Local AM Broadcast
D. Imported AM Broadcast
E. FM Sub-Carrier
G. Carrier Current
H. Local Access
I. Non-Local Access
J. Short Wave
K. Cable operated local origination service
L. Cable leased
M. Other

AUDIO SIGNAL CARRIAGE AND CHANNEL ALLOCATION CHART

Service Type:______________________________

Cable Channel/ Frequency	Program Source Licensee	Program Format	Station Call Letter or Service	Broadcast Hours Channels Per Day	Activation Date & Duration

Specify activation date if other than initial date and specify for how long this particular service will be provided.

2. Attach copies of the contracts under which applicant shall obtain the rights to programming services other than local broadcast radio signals stated in its proposal. Such contracts should indicate, at a minimum, the exact special programming services that the applicant plans to offer and distribute and the cost of such programming rights which the applicant shall pay to the owners or supplier thereof for such programming. If such programming is to be purchased on a sliding scale depending upon the number of its subscribers, indicate the cost for such programming per subscriber per year. Indicate the period during which the applicant shall be entitled to such programming rights.

3. What policies, procedures and/or practices will you adopt and implement to insure compliance with all applicable federal, state, and local laws governing obscenity?

4. Describe in detail the number of channels available for lease, including those available for full-time and part-time lease. State the terms, conditions, rates, and any other policies you will establish concerning leased channels. Attach a copy of contract forms and application forms you will use with any leased channel user.

5. Describe any special services you will make available for blind subscribers. Indicate the services, rates, and dates services will be made available.

6. Describe any special services you will make available to minority subscribers. Indicate the services, rates, and dates services will be made available.

7. Describe in detail any special services you will provide for business, industry, government, educational, or other institutional users. Specify service, rates, dates available, and duration of service.

Local Franchisee Origination Programming

If applicant plans to engage in any type of audio programming produced locally and/or distributed by the franchisee, the following items must be completed.

Note: All equipment (make, model and actual purchase price) listed by the applicant in response to this section must actually be provided. If applicant intends to use equivalent items, they must first be approved prior to substitution if the applicant is selected by the City.

1. Number and channel designation, with activation date of each channel, to be used by franchisee for locally produced and/or distributed audio programming.

- Amount of time daily each channel will be operational.
- Signal transmission quality of each franchisee local origination channel.
- Details of all stationary and mobile production facilities for franchisee local origination programming including location, size, operation schedule and operational date of each such facility.
- List all production equipment (make, model, quantity and cost) for franchisee local origination programming.
- Discuss policies and procedures governing franchisee local origination programming.
- Describe how programming needs will be determined.

2. Describe for each year of the Agreement your plans to engage in franchisee local origination programming.

3. Include a complete description of all rules, regulations, and procedures that will govern franchise local origination programming including, but not limited to, channels, equipment, and facilities.

4. Describe in detail any specific plans for dealing with any broadcast radio station located in (city, region, state).

5. Describe any special conditions, rules, or preferences attached to the use of franchise local origination equipment or facilities for Community Communications purposes, if any.

6. Detail franchisee local origination operation budget, capital budget, and staff for each year of the term of the Agreement. All financial figures should be in constant 198__ dollars.

7. Describe in detail the type of programming you will produce and/or distribute locally and the number of hours for each type of programming. How will programming needs be determined?

8. Describe in detail, for each year of the term of the Agreement, your plans and policies for programming designed to meet the needs of minority persons and women in the City. The statement should also include a description of how these needs will be ascertained.

Audio Community Communications Provisions (Access)

If applicant plans to offer any Community Audio Communications provisions in its proposal, the following items must be completed.

Note: All equipment (make, model and actual purchase price) listed in response to this section must actually be used for community communications. If applicant intends to use equivalent items they must first be approved by the Superintendent prior to substitution, if the applicant is selected by the City.

1. Applicant should state in detail all aspects of its proposed system design for Community Audio Communications, which shall include, but not be limited to, the following:

- Number and channel designation of city-wide Community Audio Communications channels and separately programmable area channels, with activation date of each channel.
- Additional channels the applicant will reserve and make available for community communications uses in the future and under what conditions such channel will be made available and when.
- Amount of time daily each Community Audio Communications channel will be operational.
- Signal transmission quality of each Community Audio Communications channel.
- Details of all stationary and mobile production facilities

that will be available to Community Audio Communications users, including location in each of the separately programmable areas, size, operating schedule, and operational date of each such center.

o Program distribution and interconnection capability for each separately programmable area and City-wide distribution.

o List all production equipment (make, model, quantity and cost) that will be made available for Community Audio Communications users.

o Policies and procedures for reserving production equipment, studio facilities, and channel time.

o Number and availability of trained personnel to assist users of production equipment and facilities.

o Detail policies and procedures governing who will be able to use each Community Audio Communications channel.

o To what extent, if any, will the applicant exercise editorial control of programming over any Community Audio Communications channel.

o Detail policies for free use of channel time and production facilities or mobile equipment.

o Describe policies governing multiple uses of any Community Audio Communications channel or production equipment by an individual user.

o List all rates to be charged for use of any Community Audio Communications channel and/or production equipment for Community Audio Communications users.

2. Discuss in detail any provisions you will have for direct audio feed capabilities from institutions (list each one by name) within the City such as, but not limited to, schools, university subheadends, public radio stations, municipal offices, churches, synagogues, hospitals, libraries, museums, governmental buildings, or other entities for program distribution over a Community Audio Communications channel.

3. For each year of the Agreement, detail the specific channels, equipment, facilities and services for Community Audio Communications usage you will provide.

4. State all responsibilities that the applicant will accept and all rules and regulations the applicant will impose on the users. Include a complete description of all rules, regulations, and procedures that will govern the proposed Community Audio Communications aspects of the cable system.

5. State the role users will have in the decisions of applicant

relative to operational and programming policies and procedures dealing with Community Audio Communications.

6. Detail the operating budget, capital budget, and staff for Community Audio Communications for each year of the Agreement. All financial figures should be in constant 198__ dollars.

7. Describe in detail damage deposit charges to be required of any user of Community Audio Communications equipment.

Technical Specifications

The RFP should include technical specifications, many of which will be included as part of the video technical specifications. Where this occurs, they should be identified as audio as well as video specifications to assure their consideration. The reason for this can best be illustrated when it is understood that many cable system technical specifications preclude delivery of sub-carriers. The following check list should be helpful:

1. Antenna
2. Converters
3. Technical Standards
4. Testing Procedures, test equipment to be used (make and model) and number and general location of test points for each test. Also a description of the method and frequency of test equipment calibration, forms, and methods of recording field data and permanent test results, recordkeeping, forms, and method of initial proof of performance testing and details of the annual proof of performance tests.
5. Procedures for routine preventative and ongoing maintenance, including type and frequency of system inspection and testing, number and qualifications of audio technical staff, and service facilities.
6. Provisions for emergency override capability of the system and emergency channel allocation.
7. Detailed description of frequency spectrum and channel capacity for audio signal carriage capability of the system; activation dates for each aspect of the system; number of fully usable channels on the FM band, on TV audio channels, and FM sub-carrier.
8. Detailed description of when, how, and under what circumstances future channel capacity will be provided to subscribers.
9. Detailed description of channel spacing.
10. Detailed description of headend shielding of FM off-air signals.

This list is not intended to be inclusive.

VIII. TEXT SERVICES

Description

As cable television expands its capabilities, text services are certain to become an inceasingly important way of distributing information quickly and inexpensively. Text services include not only repeated messages entered into the system by a character generator, but they can also include messages generated by a computer. Although computer-generated text services are not yet widely available, they are developing rapidly and are now being tried experimentally.

Text services should be of particular interest to libraries, as they have the potential of augmenting print distribution of information in a very powerful way. In the next few years, the cable industry is likely to develop text services that will allow subscribers access to current data bank information about the stock market, travel schedules and rates, store sales, etc. Some libraries are already experimenting with ways of permitting citizens access to the text services cable will be capable of delivering.

Recommended RFP Language

The city encourages applicants to provide local input capability for the text and data services they will be offering.

Questions:

Describe the local input capability the system will provide for text and data services, if any. Include type of system and capacity designated for local use (e.g., videotext, 100 pages designated for local information) and any priority that might be given to non-commercial use.

List non-commercial users identified in the ascertainment of community needs.

IX. SATELLITE ACCESS

Description

As more and more programming is available from satellites, cable operators will have to make choices about what programming will be carried on the system. Communities may wish to have programs shown on the access or LO (or other) channels that the cable operator might not normally choose to take. By having dedicated (sole) use of a receiver, the community can make its own choices concerning the programs that will be available from the satellites.

In addition to commercial program offerings that a cable operator often uses to pay to receive, satellites carry many programming services that are free or non-commercial which might interest access users. For example, access centers have carried educational programming and telecourses offered by the Appalachian Community Service Network (ACSN).

There is also the potential of access centers working with national organizations such as the Public Service Satellite Consortium to serve as teleconferencing sites. For this reason, it is very useful for the institutional network to be able to interface fully with the satellite receiver.

Recommended RFP Language

The city encourages bidders to think creatively on how satellite terminals can be used for community access and will look favorably on those who will provide services like dedicated use of a receiver by the access center at no charge.

Questions

1. Describe the access to the satellite terminal the company will provide, if any.
2. Describe the equipment that will be provided for access use and the cost for users, if any.
3. For the purpose of participating in national teleconferences, describe how many points on the system will be able to receive a satellite signal without it being broadcast to home subscribers and/or other institutions. List the organizations or institutions who will have this capability, if any.
4. Describe who will manage and control community access use of the aforementioned services, for example, the Community Access Corporation, the cable company, etc.

X. FRANCHISE FEE

Description

The franchisee fee is a sum of money, usually a designated percentage of the cable system's gross revenues, that is returned to a municipality to reimburse it for expenses incurred in issuing a franchise and regulating cable television. This money, or a sum equal to a designated portion of it, often contributes to the support of access centers and channels.

The FCC now limits the franchise fee to 3 percent of gross revenues unless a waiver is obtained to allow a commuity to receive up to 5 percent. The language presented in the recommended franchise language has been used in several FCC-approved waivers. The operator may argue for a lesser fee and question the community's ability to gain the necessary waiver in an attempt to avoid paying the increased fee. However, many operators do not object to the 5 percent fee if the city intends to use the fee to support the development of the access channels and other local cable services. In fact, the FCC waiver is usually granted only if the community will be using at least the additional 2 percent to support local programming.

The city should be very careful in developing a definition of "gross revenues" as different definitions can make a difference in thousands -- in large cities even hundreds of thousands -- of dollars over the life of the franchise. Cities should ensure that the definition of gross revenues not be limited to subscriber revenues only, but that it also include revenues derived by the cable operator from advertising, pay-per-channel and pay-per-view services, leasing charges, data services charges, etc. In communities where the franchise fee has been based on "gross subscriber revenues from basic service," cable companies have been known to reduce the level and cost of basic service, putting virtually all satellite services on a middle-level tier which, although it may cost only a few dollars a month, qualifies as a pay service and is therefore exempt from franchise fee calculations.

Some communities planning to use the franchise fee to develop access programming have made arrangements to have the company pay an "average" franchise fee each quarter or half year for the first three to five years. Based on projected cable revenues for that time period, the city receives a steady series of franchise fee payments in the first few years in order to support access on a predictable level. Projected revenues are compared with the actual revenues each year, and the succeeding franchise fee payments are adjusted to compensate for any overestimating or underestimating of the gross revenues. This system brings the city and the access center more funds in the beginning of the franchise when revenues are low and access needs to make equipment purchases and meet other initial expenses.

When examining potential language for the franchise fee section of the Ordinance, the following criteria should be included:

1. Percentage of franchise fee.
2. Caveat regarding FCC current franchise fee ceilings.
3. Use of franchise fee.
4. Payment schedule of franchise fees.
5. Right to audit by city.
6. Interest charge to be used if recomputation results in additional fees owed.
7. Non-limitation clause.
8. Definition of gross revenues (in a definition section of Ordinance).

Recommended Franchise Language

"Gross Revenues" shall mean all revenue derived directly or indirectly by the Company, its affiliates, subsidiaries, parent, and any person in which the Company has a financial interest, from providing cable television services within the city including, but not limited to, basic subscriber service monthly fees, pay cable fees, installation and reconnection fees, leased channel fees, converter rentals, studio rental, production equipment and personnel fees, and advertising revenues; provided, however, that this shall not include directly upon any subscriber or user by the commonwealth, local or other governmental unit and collected by the Company on behalf of said governmental unit.

For the reason that the streets of the city to be used by the Company in the operation of its system within the boundaries of the city are valuable public properties acquired and maintained by the city at great expense to its taxpayers, and that the grant to the Company to the said streets is a valuable property right without which the Company would be required to invest substantial capital in right-of-way costs and acquisitions, the Company shall pay to the Board an amount equal to 5 percent of Company's gross annual revenue from all sources attributable to the operations of the Company within the confines of the franchise area. In the event that the law, in the future, permits some larger basis for computing this fee, the Board shall, at its election, be entitled to collect such additional monies upon sixty (60) days prior written notice to the Company. Provided, however, that upon receipt of such notice, the Company shall be entitled to initiate a petition for an equivalent rate increse notwithstanding any other then pending or recently decided rate requests. In said event, the Board shall afford said rate increase petition a presumption of reasonableness and shall stay the effective date of said new fee until sixty (60) days after its decision on said rate petition.

a. It is the intent of the Board to use franchise fees as necessary, to defray the costs of local regulation of the Company, to support the development of access channels, and to generally encourage the development of the system.

b. This payment shall be in addition to any other tax or payment owed to the city or other taxing jurisdiction by the Company.

c. Payments due under this provision shall be payable forty-five (45) days after the end of each calendar quarter. Each such payment shall be accompanied by a report, certified by an officer of the Company, showing the basis for the computation thereof.

d. No acceptance of any payment shall be construed as an accord that the amount paid is in fact the correct amount, and all amounts paid shall be subject to audit and recomputation by the city for a period of one (1) year after receipt thereof. In the event that recomputation results in additional fees owed, such amount shall be subject to a ___ percent per annum interest charge.

e. Each payment of the franchise fee shall be accompanied by a certificate of an executive officer of the Company to the effect that the amount of the payment has been correctly determined in accordance with the provisions in this Ordinance.

f. The Board shall have the right to inspect the Company's income records and the right to audit and to recompute any amounts determined to be payable under this Ordinance; provided, however, that such audit shall take place within thirty-six (36) months following the close of each of the Company's fiscal years. Any additional amounts due to the Board as a result of the audit shall be paid within thirty (30) days following written notice to the Company by the Board which notice shall include a copy of the audit report.

XI. UNIVERSAL SERVICE

Description

Universal service refers to the concept of offering a basic level of cable service--usually a small number of access and local channels--to any household requesting such service without charging a monthly subscription fee. A one-time installation fee may or may not be charged. This service, although it is being proposed in many communities, has not yet been carried out in an existing system.

Universal service can be a potential boon to community communications. The ability of cable to deliver such services as emergency alert (e.g., civil defense), coverage of community meetings and events, and distribution of community information can create a local communications medium a community can rely on. Without universal service, however, most communities will have a significant number of homes--perhaps more than half--which will not be included in the system (the national average of homes subscribing to cable systems is approximately 55 percent.) This raises serious questions about describing a local cable system as a comprehensive community communications system if approximately half the community is not receiving the service. It is, therefore, a significant benefit to the community to have universal service be a component in its cable plan, because it has the potential to significantly increase the percentage of homes receiving community information. Also, because this service has the potential of wiring so many homes into an information/data distribution system, it can facilitate the introduction of municipal or public services such as meter reading and fire protection devices.

Cable television will increasingly be relied on as an efficient and inexpensive way of distributing information locally. If a municipal government or service agency can use cable to distribute a message rather than using print messages which are mailed to citizens' homes, they will be able to save considerable amounts of money as the cost of print messages increases and the cost of electronic messages holds steady or decreases in real dollars. If, however, cable reaches only half of a city's households, the savings are completely lost because a message will have to be duplicated in another medium, probably print.

To realize the advantages of total cable coverage, a municipality may consider ways in which the subscriber will not bear the cost of delivering free basic service to non-subscribers. It should be remembered, however, that universal service can provide benefits to the cable operator if advertising is allowed on any of the universal channels, or if any of the channels are used for marketing other cable services, so that asking the cable operator to bear universal service expenses may be reasonable.

Citizens can benefit from universal service because many can receive community information that interests them without having to pay for extended cable services which they may not want. This is especially important when the needs of low income citizens are considered, as cable has the potential to increase dramatically the amount of information that has direct impact on people's lives, such as information they use to make purchasing decisions, voting decisions, financial planning decisions, etc. The availability of this information has political and social implications. By limiting the availability of this information to only those who can afford it, two classes of citizens are created: the information rich and the information poor. Universal service offers at least a portion of this information to everyone.

Universal service also benefits the cable operator. It can introduce households which might not otherwise consider receiving cable to cable services. One way that the cable operator can use this for reaching a potential market of paying subscribers is to include local origination programming and highlights of programming on pay tiers as part of the universal service package. Also, the operator benefits by expanding the audience for sponsoring local origination programs. As a marketing strategy, universal services provides the cable operator with an expanded account list for targeting in campaigns to market pay services. In general, universal service helps the cable operator surmount initial consumer resistance to having cable installed in their homes. For a universal service consumer to upgrade to a paying subscriber, the operator need only switch (in most cases) the type of converter used on the TV set in the home.

<u>Potential Problems of Universal Service</u>

1. The community should be aware that universal service might be exploited by the cable operator to offer multiple channels filled primarily with advertising and limit the number of access or community channels.

2. The frequent argument against universal service is the uncertainty of its economic feasibility. If, for instance, universal service

is installed for free or at a rate below actual installment cost, the loss is absorbed by other paying users of the system.

3. Communities proposing this service have to be aware that this concept has not actually been field tested. There is some question as to the exact amount of increase in the percentage of potential subscribers when companies require from $25 to $50 as a one-time installment fee as in most universal service proposals. At the same time they might offer in a tiered proposal a free installation for a $3.95 a month basic service. It is not known which marketing strategy would attract more users or whether free or very low cost installation would create an unacceptable economic burden on the system.

XII. INTERCONNECTION

Description

As cities examine the issue of franchising or renegotiation, one very important but often overlooked issue is interconnection.

Interconnection is of particular importance in suburban areas. In suburban areas, school districts frequently cross municipal boundaries, or social service districts encompass groups of cities. The schools and social service agencies are providers of valuable community information, yet if there is no mechanism for them to reach their constituency (which frequently encompasses several cities or groups of cities) cable technology cannot be put to its best use.

Technically, interconnection can take place for one or more channels several ways--by actually connecting cables in adjoining municipalities, by microwave, by ITFS, by a low-power television transmitter, or by any other method of transferring a signal from one system to another. A wire or microwave connection is by far the most common.

The city may or may not choose to specify the technical method of interconnection. It should, however, specify the number and type of channels to be interconnected as well as specifying a date when interconnection must be accomplished.

Historically, interconnection has not taken place for political and commercial reasons. Unless cities require interconnection, few cable companies will go to the expense. When two or more different companies are involved, coordination can become difficult, often with no company willing to take the first step. Under these conditions, the city should assume a leading role in developing interconnection.

There are a number of examples of cost-effective interconnection. Programming varies from traditional instructional television to split screen interactive communication (for more information contact Berks Community TV, 1112 Muhlenberg Street, Reading, PA; (215) 374-3065). In Kettering, Ohio, an interconnect is designed to deliver regional hearings, instructional programs, and community information among the six cities which form the Miami Valley Cable TV Council (for more information contact Miami Valley Cable Council, 3700 Far Hills Avenue, Kettering, Ohio 45429; (513) 298-7890).

Interconnection can be financially attractive to a cable company. For instance, when two companies share an interconnect, each company typically pays 50 percent of the cost. In exchange each cable operator receives an additional channel of marketable cable service. With the cable industry's recent emphasis on diversity of programming, an additional channel or channels at no cost except 50 percent of the interconnect's capital expense and cost of ongoing maintenance can make the interconnect financially attractive to cable companies.

Recommended Franchise Language

Interconnection Required

The company shall interconnect access channels of the cable system with any or all other CATV systems in adjacent areas, upon the direction of the city. Interconnection of systems may be done by direct cable connection, microwave link, satellite, or other appropriate method. (The RFP could request that appropriate methods be described in some detail, including the applicant's proposal for interconnection, which could be included along with completion schedules.)

One channel shall be interconnected whenever any one or more interconnected channels is in use during 60 percent of the time between the hours of 9:00 a.m. to 6:00 p.m. and 80 percent of the time between the hours of 6:00 p.m. and 9:00 p.m. on weekdays (Monday through Friday); the company shall make an additional channel available for the same purposes up to the channel capacity of the system.

Interconnection Procedure

Upon receiving the directive of the city to interconnect, the franchisee shall immediately initiate negotiations with the other affected system or systems in order that all costs may be shared equally among cable companies for both construction and operation of the interconnection link.

Relief

The franchisee may be granted reasonable extensions of time to interconnect or the city may rescind its order to interconnect upon petition by the franchisee to the city. The city shall grant that request if it finds that the franchisee has negotiated in good faith and has failed to obtain an approval, or the cost of the interconnection would cause an increase in subscriber rates that the city finds unacceptable.

Cooperation Required

The franchisee shall cooperate with any interconnection corporation, regional interconnection authority, or city, county, state, and federal regulatory agency which may be hereafter established for the purpose of regulating, financing, or otherwise providing for the interconnection of cable systems beyond the boundaries of the city.

Initial Technical Requirements to Assure Future Interconnection Capability

(1) All cable systems receiving franchises to operate within the city shall use the standard frequency allocations for television signals. (2) All cable systems are required to use signal processors at the headend for each television signals. (3) The city also urges franchisees to provide local origination equipment that is compatible throughout the area so that videocassettes or videotapes can be shared by various systems.

Glossary

In using this glossary, the reader needs to be aware that occasionally some terms over-lap, occasionally some terms which originally had different meanings are now used interchangeably, and in some cases no one standard term has been formally or informally agreed upon. An attempt has been made to be as clear, accurate, and specific as possible in dealing with these problems. Also, some of the terms included here have other definitions for use in other fields; for the most part, this glossary defines the terms only as they apply to cable/broadband.

access channels: channels which cable operators make available to others. Under rules the FCC enacted in 1972, cable operators in the top 100 TV markets had to offer four categories of access channels: public, educational, government, and leased. Although the U.S. Supreme Court overturned the rules, the terms are still in use. Many cable operators voluntarily offer access channels, and some states and cities require them.

addressability: the capability of sending signals downstream from the headend to specific locations or "addresses" on the cable system. This service requires an addressable converter at the subscriber's terminal.

addressable, programmable converter/descrambler (APCD): refers to a single subscriber terminal for cable TV in which all four functions are combined.

ADI: area of dominant influence. Term used by the Arbitron ratings service to indicate the area in which a single television station can effectively deliver an advertiser's message to the majority of homes. (Nielsen's term is DMA [designated market area].)

aerial plant: cable installed on a pole line, or similar overhead structure; space often leased from the local telephone or power company. Compare with *buried plant.*

amplifier: device used to increase the power, voltage, or current of input signal. Used in a cable system's distribution plant to overcome the effects of attenuation caused by the coaxial cable.

amplitude: magnitude or range of voltage, power, etc. of an electrical signal or radio wave.

analog: transmission format that is a continuously variable signal, generally represented by a flowing wave. Compare with *digital.* To convert signals from one of these transmission formats to the other, some form of modem (modulator/demodulator) is used.

antenna: device used to aid in the reception or transmission of signals.

APCD: *addressable, programmable converter/descrambler.*

ascertainment: in broadcasting, a requirement that stations determine the needs and interests of the community, through interviews and surveys, and design programming to help meet those needs and interests. Ascertainment studies are sometimes used by cable television companies as well.

aspect ratio: ratio of width to height in television picture.

attenuation: a decrease in the amplitude of a signal as it progresses from source to receiver, such as occurs in a coaxial cable. See *amplifier.*

automated channel: a channel on a cable system dedicated for information provided in an alphanumeric or graphic format utilizing a character generator. Some examples of automated channels are time/weather, news updates, program guides, swap and shop, community calendar, travel and entertainment, financial news, consumer assistance, and real estate listings.

bandwidth: measure of the information-carrying capacity of a communication channel. The bandwidth corresponds to the highest frequency signal which can be carried by the channel.

baseband: audio and video signal prior to its conversion to a frequency more useful for transmission via a particular medium.

basic cable service: the service that cable subscribers receive for the threshold fee—including local television stations, some distant signals, and a number of non-broadcast signals, depending on the channel capacity of the system.

bird: *satellite.*

branch cable: a secondary section of cable leading from the trunk, or distribution cable, past subscribers' homes. Also known as feeder cable.

bridging: a cable television amplifier which takes a small amount of signal from the trunk, amplifies it, and feeds it to one or more feeder lines.

broadband: relative term referring to a system which carries a wide frequency range (sometimes used to refer to frequency bandwidth greater than one MHz). In a telephone-television context, telephone would be considered narrowband (3kHz) and television would be considered broadband (6 MHz). *Broadband* is also often used to refer to communications systems such as cable TV, satellite, and microwave relay systems, which carry a large number of simultaneous transmissions. Also known as *wideband.*

broadcaster's service area: the geographical area covered by a broadcast station's signal. Also see *Grade A contour* and *Grade B contour.*

broadcasting: transmission of information by electromagnetic means, intended for public reception. Compare with *narrowcasting.*

broadcasting-satellite service: defined by the 1971 World Administrative Radio Conference for Space Telecommunications (1971 WARC-ST) as a radio-communication service in which signals transmitted or retransmitted by space stations are intended for direct reception by the general public. In an appending note the definition provided that "direct reception" encompasses both individual reception and community reception. These modes were defined as follows:

individual reception: the reception of emissions from a space station in the broadcasting-satellite service by simple domestic installations and in particular those possessing small antennae.

community reception: the reception of emissions from a space station in the broadcasting-satellite service by receiving equipment, which in some cases may be complex and have antennae larger than those used for individual reception, and intended for use: by a group of the general public at one location; or through a distribution system covering a limited area.

Compare with *fixed-satellite service,* and see also *direct broadcast satellite.*

BSS: *broadcasting-satellite service.*

buried plant: cable installed under the surface of the ground. Compare with *aerial plant.*

cable: transmission medium designed to carry either electronic or digital information over conductive or optical lines.

cable radio: broadcast and non-broadcast audio programming distributed by cable TV systems. The service predominates in areas where few FM radio signals are available over-the-air.

"cable-ready" television set: an "improved" TV set which has adequate shielding to allow use in strong local fields, and whose tuner will tune cable channels as well as standard over-the-air broadcast channels. In cable systems offering multiple pay channels, where the subscriber may choose one or more and may start and stop any of them whenever he wishes, the so-called cable-ready TV set is no longer cable-ready. The newest means of handling such services is the addressable programmable converter/descrambler (APCD). (A nationally standardized system of coding and addressing could allow these functions to be built into the subscriber's set along with cable-ready tuning.)

cable television channel: a transmission path which is used by a cable television system to deliver a signal to or from subscribers. See *Class I, II, III,* and *IV cable television channels.*

cable television system: a broadband communications system, capable of delivering multiple channels of entertainment programming and non-entertainment information, generally by coaxial cable. Many cable TV designs integrate microwave and satellite links into their overall design, and some now include optical fibers as well. Technically, for FCC purposes, a cable television system is "a non-broadcast facility consisting of a set of transmission paths and associated signal generation, reception, and control equipment, under common ownership and control, that distributes or is designed to distribute to subscribers the signals of one or more television broadcast stations, but such term shall not include (1) any such facility that serves fewer than 50 subscribers, or (2) any such facility that serves or will serve only subscribers in one or more multiple dwellings under common ownership, control, or management." Sometimes called cable system or *CATV.*

cablecasting: programming other than broadcast signals carried on cable television systems. See *Class I, II, III,* and *IV cable television channels.* Also see *local programming, local origination, origination cablecasting,* and *access channels.*

carrier: an electromagnetic wave some characteristic of which is varied in order to convey information, and which is transmitted at a specified frequency. For example, a television signal includes a video carrier and an audio carrier within the 6 MHz channel bandwidth.

CARS: *Community Antenna Relay Service.*

cascade: the operation of two or more devices (such as amplifiers in a cable television system) in sequence, in which the output of one device feeds the input of the next, thus allowing equally strong signals to reach all portions of the system.

CATV: Community Antenna Television, a term used to refer to cable television. See *cable television system.*

Certificate of Compliance: under an FCC regulatory scheme in effect from 1972 to 1978, cable system operators had to apply for and receive Certificates of Compliance prior to the commencement of service or the addition of signals. Since elimination of the requirement in 1978, cable operators need file only Registration Statements when they commence operations or add signals, and need not await further FCC action.

channel: a signal path for conveying information.

channel capacity: in a cable TV system, the number of channels that can be simultaneously carried on the system. Generally defined in terms of 6 MHz (television bandwidth) channels.

character generator: electronic device which generates letters, numbers, or symbols, directly on a television screen. Usually used to supply paginated information, such as news, time, weather, etc., for display on subscribers' TV receivers.

churn: rate of turnover in subscriptions to cable and/or pay television service.

Class I cable television channel: as defined by the FCC, a signalling path provided by a cable television system to relay to subscriber terminals television broadcast programs that are received off-the-air or are obtained by microwave or by direct connection to a television broadcast station.

Class II cable television channel: as defined by the FCC, a signalling path provided by a cable television system to deliver to subscriber terminals television signals that are intended for reception by a television broadcast receiver without the use of an auxiliary decoding device and which signals are not involved in a broadcast transmission path.

Class III cable television channel: as defined by the FCC, a signalling path provided by a cable television system to deliver to subscriber terminals signals that are intended for reception by equipment other than a television broadcast receiver or by a television broadcast receiver only when used with auxiliary decoding equipment.

Class IV cable television channel: as defined by the FCC, a signalling path provided by a cable television system to transmit signals of any type from a subscriber terminal to another point in the cable television system.

closed circuit: any communication transmission method by which reception is not available to the general public. The receiving equipment is directly linked to the originating equipment by cable, microwave relay, satellite, or telephone lines. Extensively used for monitoring in hospitals, police stations, prisons, industrial training and education programs, schools, etc; for long-distance programming to specialized audiences, such as for certain sporting events and political fundraisers; and for teleconferencing.

coaxial cable: type of cable in which the center conductor is surrounded by an insulator in turn surrounded by an outer conductor. Various types of coaxial cable are the most common transmission media in cable television systems.

common carrier: a company which holds itself out to the public as providing a communications service for hire. Included in the definition are companies which own communications satellites, point-to-point microwave facilities, and MDS, as well as telephone and telegraph companies. Some have argued that cable television should be deemed a common carrier, at least insofar as the cable company leases channels.

Communications Act of 1934: federal statute which replaced the Radio Act of 1927 and established the FCC. The purpose of the Act is to "make available, so far as possible, to all the people of the United States, a rapid, efficient, nation-wide, and world-wide wire and radio communications service."

Community Antenna Relay Service (CARS): the microwave radio relay service reserved by the FCC for transmitting programming from one point to another by cable TV operators.

compulsory license: legal requirement, under the Copyright Law, for copyright holder to license users of their copyrighted material on a uniform basis and for a stipulated fee. Under this system, cable operators make payment for certain of the broadcast programming they carry, to the Copyright Office, which through its Copyright Royalty Tribunal distributes the payment to copyright holders.

conductor: any material, such as wire or coaxial cable, capable of carrying an electric current.

conduit: thin metal or plastic pipe used for protecting wire or cable.

connector: mechanical or electrical device that is used to join two or more wires, cables, circuits, or components.

contour: predicted, theoretical coverage area of a broadcast station. Under the FCC rules, TV stations have three contours: City Grade, Grade A, and Grade B—each respectively covering a wider area, but each of lesser signal strength.

converter: device for changing the frequency of a television signal. A cable headend converter changes signals from frequencies at which they are broadcast to clear channels (and from UHF to VHF since cable cannot carry UHF signals). A subscriber converter ("set-top converter") extends the channel capacity of the home television receiver. The set-top converter will have either buttons or dial, for the subscriber to select channels.

cross-modulation: see *intermodulation distortion.*

cross-over: spectrum between that used for upstream transmission and that used for downstream transmission in a two-way cable system. Also see *sub band.*

cycle: one complete alternation of a sound or radio wave. The rate of repetition of cycles is the frequency. Also see *hertz.*

DBS: *direct-broadcast satellite.*

decoder: electronic device which translates signals in such a way as to recover the original message or signal. Compare with *encoder.*

dedicated channel: a cable television channel solely used for a particular type of service, such as education, police, meter reading, public access, library, business data services, etc.

digital: a transmission format that involves pulses of discrete or discontinuous signals. Compare with *analog.* To convert signals from one of these transmission formats to the other, some form of modem (modulator/demodulator) is used.

direct broadcast satellite: satellite designed to transmit signals intended for direct reception by the general public. See *broadcasting-satellite service.*

discrete address: transmission from one single point to another single point. Compare with *multiple address.* And see *point-to-point service.*

dish: *earth station.*

display: the visual information shown on a television receiver.

distant signal: signal of a television station which would not be deemed "local" to a given cable television system under FCC's rules. Generally, signal originating at a point too distant to be picked up by ordinary television reception equipment.

distribution system: collective term for the part of the cable television system used to carry signals from the headend to subscriber TV receivers.

DMA: designated market area. Term used by the Nielsen ratings service to indicate the area in which a single television station can effectively deliver an advertiser's message to the majority of homes. (Arbitron's term is ADI [area of dominant influence].)

domsat: abbreviation for domestic (U.S.) satellite.

downlink: satellite-to-ground transmission. Compare with *uplink.*

downstream: in a cable system the direction from the headend to the subscriber terminals. Compare with *uplink.*

drop cable: that cable which feeds the signal to an individual customer from the feeder cable serving the specific area.

dual trunk capability: see *multiple cable system.*

earth station: dish-shaped antenna used for the reception (and if designed for it, transmission) of satellite signals. Sometimes called *dish.*

education access channel: a channel on a cable system dedicated for use by educational entities, generally offered free of charge by the cable system. Also see *access channels.*

electromagnectic spectrum: the continuum of frequencies generally useful for transmission of information or power by electromagnectic means. Differing bands of frequencies are allocated to different types of communications services.

encoder: electronic device that breaks up a signal into component parts for transmission of that signal. Compare with *decoder.*

exclusivity: the sole right to air a program within a given period of time in a given market.

facsimile (FAX, FX): system of telecommunications for the transmission of printed or graphic material, converted into electronic signals at the source and carried to the subscriber's receive terminal where it is reconverted into a copy of the original.

feeder cable: a secondary section of cable leading from the trunk, or distribution cable, past subscribers' homes. Also known as branch cable.

fiber optics: see *optical fiber.*

field: one half of a television picture—the odd or even scanning lines. Two fields are interlaced to form one frame or complete picture, and scanning occurs at 60 fields per second. Also see *frame* and *scanning line.*

filter: circuit within a cable distribution system that allows passage of desired channels and blocks others, such as for distribution of different packages of service.

fixed-satellite service: defined by the 1971 World Administrative Radio Conference for Space Telecommunications (1971 WARC-ST) as a radio-communication service between earth stations at specified fixed points when one or more satellites are used, in some cases including satellite-to-satellite links; and for connection between one or more earth stations at specified fixed points and satellites used for a service other than the fixed-satellite service (for example, the broadcasting-satellite service, etc.). Compare with *broadcasting-satellite service.*

footprint: satellite coverage area.

frame: one complete television picture consisting of two fields of interlaced scanning lines. A frame lasts 1/30 of a second.

frame grabber: control device allowing an individual television viewer to stop a single frame of a moving TV picture and hold that frame for as long as the viewer may wish to see it.

franchise: an authorization or license issued by a political subdivision for the operation of a cable television system. The document usually sets out the specific rights and responsibilities of each party to the agreement. See *ordinance.*

franchise fee: annual fee collected by the franchising authority from the cable operator, generally based on a percentage of the cable operator's gross revenues, and used to cover such things as cost of use of the public right-of-way, regulatory activities, and other cable-related expenses. FCC regulations limit the allowable franchise fee to three percent of gross revenues, or five percent with special permission.

frequency: the number of complete alternations of a sound or radio wave in a second. See *hertz, kilohertz, megahertz,* and *gigahertz.*

frequency-division multiplex: technique by which two signals are transmitted simultaneously on the same transmission medium; each signal is assigned to a separate and distinct carrier frequency within the medium. Also see *time-division multiplex.*

FSS: *fixed-satellite service.*

geostationary orbit (GSO): a special class of geosynchronous orbit; a circular equatorial orbit in which a satellite appears to remain stationary with respect to the surface of the earth. Note that the acronym *GSO* is used to refer to both geostationary orbit and geosynchronous orbit. Also see *orbital position.*

geosynchronous orbit (GSO): a satellite orbit with a period of exactly one day. Also see *geostationary orbit* and *orbital position.*

geostationary satellite: orbital communications satellite moving at the speed of the earth's rotation, thus apparently stationary as viewed from the earth's surface. Also known as synchronous or geosynchronous satellite.

geosynchronous satellite: see *geostationary satellite.*

gigahertz (GHz): a unit of frequency equivalent to one billion hertz or cycles per second.

Grade A contour: geographical reception area of a broadcast station wherein satisfactory reception is estimated to be available 90 percent of the time at 70 percent of the receiver locations. Part of the broadcaster's service area. Also see *Grade B contour.*

Grade B contour: geographical reception area of a broadcast station wherein satisfactory reception is estimated to be available 90 percent of the time to 50 percent of the receiver locations. Often spoken of as 35 miles, but actually could be far more or less, depending on terrain factors and on antenna efficiencies at the transmitter.

grandfathering: generally applies to situations in which government changes a policy but allows persons doing the thing newly-prohibited to continue doing it.

GSO: acronym used interchangeably to refer to (1) *geosynchronous orbit* and (2) *geostationary orbit.*

hardware: actual physical equipment, such as cameras, recorders, etc., as distinguished from materials like programming. Compare with *software.*

HDTV: *high-definition television.*

headend: control center of a cable television system, where incoming signals are amplified, converted, and combined in a common cable along with any origination cablecasting, for sending out to subscribers.

hertz (Hz): a unit of frequency equivalent to one cycle per second. See *cycle.*

high definition television (HDTV): any of several improved forms of television photography, recording, transmission, and reception. Improvements may include higher resolution, higher sound fidelity, (generally stereo), more faithful color rendition, and a wider aspect ratio.

home information utility: a service which involves the delivery of "alpha-numeric' material (words and numbers)—and sometimes video and audio as well—to a TV screen. It comes in a variety of forms (see *videotex*, for example). By using a device at home, viewers can request information to be displayed on the TV screen. The home device may be as simple as a pad which resembles a hand-held calculator; or it may be a computer-keyboard which permits complicated interactive "conversations" with the system's computer. The data in the system may be as simple as airline timetables, stock market figures, or recipes; or it can be as complicated as financial planning, budgeting, and tax-return calculations.

homeset: in direct broadcast satellite (DBS) service, refers to the equipment necessary to receive signals—consisting of a dish antenna about three feet in diameter mounted on the rooftop of a single family home, and the electronics necessary to convert and descramble the signals from the satellite.

hub-network: a modified tree-network in which signals are transmitted to subordinate distribution points (hubs) from which the signals are further distributed to subscribers. Also see *tree-network.*

independent station: a commercial television broadcast station that is not affiliated with one of the three major commercial television networks.

infomercial: program-length commercial message. Each advertiser could create its own infomercial or the cable operator could develop special shoppers' programs by grouping commercial messages together. And the cable operator could group infomercials around a special theme, similar to a newspaper's "Home" or "Wedding" sections, for example.

institutional network: in a cable system, a dedicated network (sometimes separate, sometimes integrated with the general subscriber network) for use by institutions such as schools, hospitals, government and non-profit agencies, and business. Sometimes called *special service network.*

Instructional Television (ITV): a television system used primarily for formalized instruction.

Instructional Television Fixed Service (ITFS): the frequencies set aside by the FCC for use by educational institutions in relaying ITV programs.

interactive cable system: see *two-way cable.*

interconnection: use of microwave, satellite, coaxial cable, optical fiber, or other apparatus or equipment for the transmission and distribution of signals between two or more cable systems for mutual distribution of programming.

interference: extraneous disturbance that causes degradation or disruption of normal signal transmission.

intermodulation distortion: form of interference involving the generation of interferring frequencies during signal processing.

ITFS: *Instructional Television Fixed Service.*

ITV: *Instructional Television.*

kilohertz(kHz): a unit of frequency equivalent to one thousand hertz or cycles per second.

laser: from Light Amplification by Stimulated Emission of Radiation. Device for transmitting light in coherent form.

leased access: a channel on a cable system dedicated for lease to seaprate program providers; usually utilized by for-profit entities. Also see *access channels.*

local distribution service (LDS): a communications system used for point-to-multipointservice in a local area. The technology used may be telephone lines, microwave, or coaxial cable.

local government access channel: a channel on a cable system dedicated for use by local government and municipal entities, generally offered free of charge by the cable system. Also see *access channels.*

local loop service: historically, telephone communications path used to transmit telephone and data communications within a community. Term now used also to refer to local distribution of data, whether by telephone circuits, cable TV, or other technlogy.

local origination: term usually used for programming produced by the local cable operator. It may also be film or videotape produced elsewhere and sold or leased to the operator. Term is sometimes used interchangeably with *local programming.* Also see *origination cablecasting* and *access channels.*

local programming: collective term which may encompass not only cable system local origination, but also public, government, and educational access programming. Term is sometimes used interchangeably with *local origination.* Also see *access channels.*

local signals: television signals received at locations within a broadcaster's service area. Term also sometimes used interchangeably with *must-carry signals* which are required of cable systems by the FCC.

low power television (LPTV) station: a type of station with greatly reduced power compared with a conventional station, resulting in a much smaller Grade B contour and broadcast service area; it is licensed by the FCC in a manner similar to a television translator station, except that an LPTV station may originate programming. LPTV stations area not classified by the FCC as must-carry signals for cable systems.

LPTV: *low power television station.*

Major Markets: most often refers to the top 100 television markets in the United States. Sometimes used to refer to the five or ten or so of the very largest television markets in the country. Also see *Top 100 Markets.*

mandatory carriage: television broadcast signals that a cable system must carry in accordance with FCC regulations. See also *signal carriage rules.*

master antenna television (MATV): coaxial cable distribution system of signals received by a master antenna, serving one building or adjacent groups of buildings (such as apartments, hotels, or hospitals). When a satellite dish is used in conjuction with the master antenna, the service is often referred to as satellite master antenna television (SMATV), or mini-cable. MATV and SMATV systems are distinguished from cable television systems by the absence of any local franchise or regulation. Such regulation is avoided by not placing cables over or under public streets or rights of way. Such a system offering a full range of services may be a cable television system for purposes of regulation by the FCC or for purposes of the Copyright Law if more than 50 subscribers are served and the buildings served are not commonly owned, managed, or controlled.

MATV: *master antenna television service.*

maxi-pay service: full service pay programming; from eight to twenty-four hours a day of continuous programming that includes first-run movies, plus, in many cases, sports specials and entertainment features. Compare with *mini-pay service.*

MDS: *multipoint distribution service.*

megahertz (MHz): a unit of frequency equivalent to one million hertz or cycles per second.

microwave link: a relay station in a microwave relay system that receives, amplifies, and retransmits signals. See *microwave relay system.*

microwave relay system: line of sight, point-to-point transmission of television and other signals at high frequencies, by means of geographically spaced microwave links.

microwaves: the radio fréquencies above 1000 megahertz. Used in point-to-point communications.

Midwest Video I: refers to the court challenge to the FCC's 1969 action requiring cable systems with 3500 or more subscribers to originate programming. In 1970 Midwest Video Corporation appealed the action. The company obtained a jucidial stay of the mandatory aspects of the rules and successfully prosecuted its appeal before the U.S. Court of Appeals, Eighth Circuit. The FCC took the matter to the Supreme Court and in 1972 a closely divided Court upheld the FCC's authority, but the FCC abandoned the rule two years later. Later, Midwest Video Corporation was to challenge the FCC's cable access and channel capacity rules and usage of the term *Midwest Video* now usually has reference to the more recent case; see *Midwest Video II.*

Midwest Video II: refers to the court challenge to the FCC's rules on cable access and channel capacity brought by Midwest Video Corporation. In February 1978, the U.S. Court of Appeals, Eighth Circuit, in *Midwest Video Corporation* vs. *FCC,* set aside these rules as exceeding the FCC's jurisdiction. In April 1979, the Supreme Court, in *FCC* vs. *Midwest Video Corporation*, affirmed the Eighth Circuit's decision. See also *Midwest Video I.*

mini-pay service: pay programming available for a limited number of hours a day; cost of this service to subscribers is about half that of the maxi, or full, service. Compare with *maxi-pay service*.

modem: modulator/demodulator device, such as is used, for example, to connect a home computer to an ordinary home telephone. Also see *modulation*.

modulation: process whereby original information can be translated and transferred from one medium to another, each capable of duplicating the pattern of amplitude and frequency of which the signal consists. Also see *modem*.

monitor: a video display unit having no radio frequency tuning capability. A monitor may be used, for example, to view a TV picture directly from a TV camera.

MSO: *multiple system operator.*

multiple address: transmission from one single point to a number of selected specific points. Compare with *discrete address*. And see also *point-to-point service.*

multiplexing: the combining of two or more signals into a single transmission from which the signals can be individually recovered.

multipoint distribution service (MDS): a common carrier microwave radio service authorized to transmit private television and other communications. MDS provides an omnidirectional signal in the 2150-2162 frequency range, and the service may carry up to two full video channels, depending on the market. The service has proved to be an effective means of delivering pay TV programming especially to apartment buildings and hotels. Pending before the FCC (as of October 1982) is a proposal to reallocate a portion of the 2500-2690 MHz band to make more channels available for MDS, thereby making provision for multi-channel MDS systems.

multiple cable system: a system using two or more cables in parallel to increase information-carrying capacity.

multiple system operator (MSO): an organization that operates more than one cable television system.

must-carry signals: those broadcast signals which the FCC requires a cable system to carry; based on fixed mileage zones, audience surveys, market location, and TV stations' signal contours, in relation to the particular cable TV system's service area.

narrowband: a relative term referring to a system which carries a narrow frequency range (sometimes used to refer to frequency bandwidths below one MHz). In a telephone-television context, telephone would be considered narrowband (3 kHz) and television would be considered broadband(6 MHz).

narrowcasting: transmission of information by electromagnetic means, intended for a particularly audience (for example, industrial TV, special-audience cable TV, and business and professional programming). Compare with *broadcasting*.

network: a national, regional, or state organization that distributes programs to broadcasting stations or cable television systems, generally by interconnection facilities. The term has generally referred to the three major television networks, but by 1980 many new networks had emerged—commercial, broadcasting and cablecasting, pay TV, religious programming, etc.—such development greatly facilitated by the availability of satellite transmission.

noise: the accidental, unintended, and normally unwanted components of information received or transmitted as electrical impulses. These interferences degrade the transmission of the desired signal (for example, "snow" in a television picture).

non-broadcast channels: see *cablecasting*.

nonduplication: usually refers to the FCC's network program nonduplication rules, which require cable operators to protect the network programming broadcast by local affiliates by blocking out the programming carried simultaneously by distant stations which the system also carries. This rule still exists, although the term "nonduplication" sometimes also refers to the FCC's syndicated program exclusivity rules which the FCC has eliminated.

optical fiber: thin fiber of very pure glass highly transparent, used for transmitting information by means of light. There is some use of optical fibers in cable television systems today for specific purposes and their use may become more common in the future. Optical fibers have potentially a very high capacity for carrying information, but this is not realized in practice at present.

orbital position: location of a geostationary satellite in orbit at a fixed point in relation to the earth. Used interchangeably with *orbital slot*.

orbital slot: see *orbital position*.

ordinance: law enacted by cities, towns, villages, and other local governmental entities. For cable television, local governmental entities usually first enact "enabling" ordinances, which set forth the terms under which applications will be submitted and a franchisee chosen, as well as general terms of service. The second step is the "granting" ordinance, usually called a franchise, which grants authority to an applicant to operate the system and sets forth specific conditions.

origination cablecasting: as defined by the FCC, programming, exclusive of broadcast signals, carried on a cable TV system over one or more channels and subject to the exclusive control of the cable operator. This is the programming on cable to which the FCC applies the fairness doctrine and "equal time" rules.

pay cable: pay television programs distributed on a cable television system and paid for at additional charge above the monthly cable subscription fee. Fee may be levied on several bases: per-program, full service, tiered service, etc. See also *pay television*, *maxi-pay service*, and *mini-pay service*.

pay television: a system of distributing television programming either over the air, by MDS, or by cable, for which the subscriber pays a fee. The signals for such programming are scrambled to keep non-paying persons from receiving service, and a decoder is used to allow the paying subscribers to receive the pay television programming. Sometimes called Premium TV. See also *pay cable* and *Subscription TV*.

penetration: in areas where cable TV is available, the percentage of households subscribing to the service. Also known as saturation.

pirating: 1)making copies of copyrighted material for sale without a license from the copyright holder to do so; 2)receiving for any purpose pay-telecommunications services without making payment.

plant: the physical equipment, buildings, etc., of a broadcast station or cable system.

point-to-point service: the transmission of a signal (audio, video, or data information) via a technology such as microwave, cable, or satellite, directly to the desired receiver(s) rather than to the general public. The service can be in the form of either discrete address, from one single point to another single point, or multiple address, from one single point to a number of selected specific points.

pole attachment rights: the rights obtained by cable TV systems to attach cables to poles owned by telephone or power companies.

portapak: a relatively inexpensive, portable, battery-operated videotape recorder and camera ensemble.

premium radio: pay-radio service, which in 1982 was just in the beginning stages of operation.

premium TV: see *pay television*.

prime time: generally, the broadcast period(s) viewed by the most people, and for which a broadcast station charges the most for advertising time. As defined by the FCC, the five-hour period from 6 to 11 p.m., local time, except that in the Central Time Zone the relevant period shall be between the hours of 5 and 10 p.m., and in the Mountain Time Zone each station shall elect whether the period shall be 6 to 11 p.m. or 5 to 10 p.m.

production: the preparation and recording of a program for broadcast or cable transmission.

public access channel: a channel on a cable system dedicated for use by the public on a non-discriminatory basis, usually with no charge for channel time. Also see *access channels*.

raster: the scanned (illuminated) area of a television picture tube.

receiver: electronic device which can convert electromagnectic waves into either visual or aural signals, or both. For cable TV, usually the subscriber's television set.

redundant cable: the unused cable(s) in a multiple cable system, capped off and reserved for future use as greater channel capacity is needed or as a back up should the need occur.

regional channel: a cable TV channel dedicated for regional programming, usually involving interconnected cable systems.

Registration Statement: in cable television a filing with the FCC, which provides authority to carry television signals if such carriage is consistent with FCC rules. Replaces Certificate of Compliance.

retrofitting: the adding of additional equipment to or rebuilding sections of a cable distribution system after it has been installed. Sometimes needed to increase channel capacity or to be able to provide interactive service.

satellite: orbiting space station primarily used to relay signals from one point on the earth's surface to another.

satellite master antenna television (SMATV): see *master antenna television system.*

saturation: in areas where cable TV is available, the percentage of households subscribing to the service. Also known as *penetration.*

scanning: process of breaking down an image into a series of elements or groups of elements representing light values and transmitting this information in time sequence.

scanning line: a single continuous narrow strip of the picture area of a television tube containing highlights, shadows, and halftones, determined by the process of scanning. Also see *field.*

scramble: to break up an electronic signal into its various component parts. In pay television, for example, the signal is scrambled, and a decoder is necessary for the signal to be unscrambled so that only subscribers would receive the proper signal. A given scrambling encoder of course requires a compatible unscrambling decoder for the system to work.

side band: a band of frequencies adjacent to the carrier frequency, which is generated by modulation of the carrier, and which carries the desired information to be transmitted.

signal-to-noise ratio: the ratio of power level of the desired signal (especially television signal) to undesired noise power level present in the signal.

signal: (1) the message to be transmitted; (2) the electric impulse derived from and converted to the message being transmitted, whether audio, video, text, remote control, or other information.

signal carriage rules: the FCC rules covering the carriage of television broadcast signals on cable television systems. The FCC has no signal carriage rules for cable carriage of radio stations. Also, the FCC preempts states and cities from regulating cable carriage of broadcast signals.

slow scan television: television transmission for transmitting still pictures at a slow rate, using telephone lines or other channels having limited information-carrying capacity. Some uses are for surveillance and for graphics transmission.

SMATV: *satellite master antenna television.* See *master antenna television system.*

software: the working materials from which a program is created that will be played out, such as script, audio or visual aids, etc., especially created for the program, and knowledge of how to use the equipment to produce the program. Compare with *hardware*.

specialty station: as defined by the FCC, a commercial television broadcast station that generally carries foreign-language, religious, and/or automated programming in one-third of the hours of and average broadcast week and one-third of weekly prime time hours.

star-configuration: design for a telecommunications system in which signals from the central source are generally transmitted directly to each subscriber (as distinguished from a tree-configuration with trunk and branches to reach the subscriber). The telephone network is an example of a star-configuration.

STV: *subscription TV.*

sub band: portion of the frequency band (5 MHz to 50 MHz) often used for upstream transmission on a two-way cable system. Most subscriber-service networks use the spectrum between 5 MHz and 30 MHz for upstream transmission, with the spectrum between 30 MHz and 50 MHz, called "cross-over," unused because filters with perfect bypass characteristics cannot be built. (The spectrum from 50 MHz to the upper transmission limit would be designated for downstream transmission.)

subcarrier: a carrier which is itself imposed upon another carrier. Independent television, telephone, or other signals can be carried on different subcarriers imposed upon the same carrier.

Subscription TV (STV): form of pay service delivered over the air by scrambling a television broadcast signal.

superstations: independent television broadcast stations whose signals are available to cable systems throughout the country by satellite.

special service network: see *institutional network*.

switched system: a communications system (such as a telephone system) in which arbitrary pairs or sets of terminals can be connected together by means of switched communications lines.

synchronous satellite: see *geostationary satellite*.

syndicated program exclusivity rules: the FCC rules, now eliminated, which required cable operators to black out programs carried on distant television broadcast stations, under certain circumstances, to protect program syndicators' contractual exclusivity with local broadcasters.

syndicated programming: according to the FCC, any program sold, licensed, distributed, or offered to television station licensees in more than one market within the United States for non-interconnected (i.e., non-network television broadcast exhibition. More broadly speaking, any television programming not distributed by one of the major television networks.

teleconferencing: real-time "meeting" of individuals or groups in two or more locations, via video and/or audio hookups, Usually signals are transmitted via satellite and/or telephone lines; some teleconferencing takes place on a few interactive cable TV systems. (The term is sometimes used to encompass interactive telecommunications through computer as well.)

teletext: see *videotex.*

television channel: the range, or *band,* of the radio frequency spectrum assigned to a TV station; the U.S. standard bandwidth is 6 megahertz.

television translator station: a low-powered FM or television station receiving broadcast signals and retransmitting them on a new frequency. Authorized by the FCC especially for difficult geographical locations.

terminal: (1) generally, connection point of equipment, power, or signal; (2) usually referred to in cable TV as the point of connection between a cable drop and a subscriber's receiver, although the term is also used for any "terminating" piece of equipment such as a computer terminal.

tiered service: term used to refer to different packages of programs and services on cable TV systems for different prices; a marketing approach that divides services into more levels than simply "basic" service and one or two "pay" services. Also see *pay cable* and *basic cable service.*

time-division multiplex: technique by which two signals are transmitted on the same transmission medium; each signal is assigned to a separate and distinct time slot within the medium. Also see *frequency-division multiplex.*

Top 100 Markets (Major Markets): for regulatory purposes, the one hundred largest television markets as defined by the FCC. For advertising and other purposes, the one hundred largest television markets as defined by the ratings services.

transmission: the sending of information (signals) from one point to another.

transponder: one of several units, or "addresses," on a communications satellite capable of receiving and transmitting a full video signal. Some users lease the right from the satellite operator to use an entire transponder; others lease only a part of a transponder's capacity (such as for data communications).

translator: see *television translator station.*

trap: a device for removing a set of frequencies from a specified band of frequencies. A *positive trap* removes an interfering signal that has been intentionally introduced into the signal in order to scramble it to prohibit unauthorized reception. A *negative trap* removes a television signal itself, to prohibit unauthorized reception.

tree-network: a design for a cable system in which signals are disseminated from a central source. The configuration resembles that of a tree, in which the product from the root (headend) is carried through the trunk and then through the branches (feeders) to the individual stems (drops) which feed each individual leaf (terminal). Also see *hub-network.*

trunk: the main distribution line leading from the headend of the cable television system to the various areas where feeder lines are attached to distribute signals to subscribers.

turn-key: usually refers to installation where for a fee a contractor builds everything necessary for a complete system.

TVRO: Television Receive-Only Earth Station. See *earth station.*

twisted pair: the wiring used by the telephone system, with a capacity roughly limited to voice and low- and medium-speed data.

two-way cable system: a cable system capable of carrying information both downstream from the headend to any subscriber's terminal and from any subscriber back to the headend. The information transmitted could be of varied forms, such as audio, digital, video, or combinations thereof. Depending on the construction and sophistication required in a given cable system, the system can operate on either one or multiple lines. Also known as *interactive cable.*

UHF channels: the ultra high frequency part of the spectrum allocated for television broadcasting, comprising channels 14 through 83.

uplink: ground-to-satellite transmission. Compare with *downlink*.

upstream: in a cable system the direction from the subscriber terminals to the headend. Compare with *downstream*.

VCR: *videocassette recorder.*

vertical blanking interval: the unused lines in each frame of a television signal (which can be seen as a thick band when the TV picture rolls over) usually at the beginning of each field, which instruct the TV receiver for reception of the picture. Some of the lines can be used for teletext and captioning.

VHF channels: the very high frequency part of the spectrum allocated for television broadcasting, comprising channels 2 through 13. (The VHF band also includes the entire FM band.)

videocassette: videotape in container that provides automatic threading (tape moves in a continuous loop, rather than reel-to-reel).

videocassette recorder (VCR): electromechanical device used to record television sound and picture on magnetic-coated tape in a container that provides automatic threading, for playback on a television receiver or monitor.

videodisc: phonographic record-like device used for playing back prerecorded video (with sound) programming.

videotape: plastic tape with magnetic coating, used to record (and rerecord) and playback video and audio signals.

videotape recorder (VTR): electromechanical device used to record television sound and picture on magnetic-coated tape for playback on a television receiver or monitor.

videotex: the generic term used to refer to a system(s) for the delivery of computer-generated data into the home, usually using the television set as the display device. Some of the more often used specific terms are "viewdata" for telephone-based systems (narrowband interactive systems); "broadcast teletext" for broadcast systems (with digital information stored in the vertical blanking interval, that is, the unused portion, of the television signal); and "wideband broadcast" or "cabletext" systems (utilizing a full video channel for information transmission); and "wideband two-way teletext" (which could be implemented over two-way cable TV systems). In addition to the systems mentioned here, hybrids and other transmission technologies (such as satellite) could be used for delivery of videotex services on a national scale. See also *home information utility.*

VTR: *videotape recorder.*

wideband: see *broadband.*

Resources

Publications

Baer, Walter S. *Cable Television: A Handbook for Decisionmaking.* Report No. R-1134-NSF. Santa Monica, CA: The Rand Corporation, 1973.

Baldwin, Thomas and McVoy, D. Stevens. *Cable Communications.* Englewood Cliffs, N.J.: Prentice-Hall, 1982.

Baldwin, Thomas F. et al. *Michigan State University—Rockford Two Way Cable Project. System Design, Application Experiments and Public Policy Issues. Final Report, Volume 2.* East Lansing: Michigan State University, 1978.

Bender, Eileen T. et al., ed. *Cable TV: Guide to Public Access.* South Bend, Indiana: University of Indiana, 1979.

Bloom, L.R.; Hanson, A.G.; Linfield, R.F.; and Wortendyke, D.R. *Videotex Systems and Services.* NTIA/Report-8—50. c/o National Telecommunications and Information Administration, Institute for Telecommunications Sciences, 1980.

Bretz, Rudy. *Handbook for Producing Educational and Public-Access Programs for Cable Television.* Englewood Cliffs, N.J.: Educational Technology Publications, 1976.

Cablemark Probe, Volume I, No. I, Summer, 1982. The ELRA Group, Inc. Write: P.O. Box 70, East Lansing, Michigan 48823.

"Cable Television: Franchising and Refranchising for a Wired Community," Advisory Report produced by the Product Information Network Division of the McGraw-Hill Information Systems Company. Write: Kenneth Coughlin, Editor, for Product Information Network Division, McGraw-Hill Information Systems Company, 1221 Avenue of the Americas, New York, New York 10020.

Cowan, Robert A. *The Design, Construction and Implementation of the Interactive Telecommunications Systems for Central Maine.* Augusta, Maine: Medical Care Development, Inc., 1979.

Curtis, John A., and Pence, Jr., Clifford H. "Cable Television: A Useful Tool for the Delivery of Education and Social Services." In *Educational Telecommunications Delivery Systems.* Washington, D.C.: American Society for Engineering Education, 1980.

Easton, K.J. *Thirty Years in Cable TV: Reminiscences of a Pioneer.* Pioneer Publications, 1980.

Eilber, Carol Brown. *Cable Television: What Educators Need to Know.* Washington, D.C.: NFLCP, 1982.

Federal Communications Commission. *Information Bulletin on Cable Television.* Washington, D.C.: FCC, 1980.

Forbes, D., and Layng, Sanderson. *The New Communicators: A Guide to Community Programming.* Washington, D.C.: Communications Press, 1977.

Haight, Timothy R., ed. *Telecommunications Policy and the Citizen.* New York: Praeger Special Studies, 1979.

Hamburg, Morton I. *All About Cable: Legal and Business Aspects of Cable and Pay Television.* New York: Law Journal Seminars Press, 1982. (2nd edition)

Hollowell, Mary Louise, ed. *The Cable/Broadband Communications Book, Volume I, 1977-78.* Washington, D.C.: Communications Press, 1979.

Hollowell, Mary Louise, ed. *The Cable/Broadband Communications Book, Volume 2, 1980-81.* Washington, D.C.: Communications Press, 1980.

Jesuale, Nancy, ed., with Smith, Ralph Lee. *Cable Television Information Center Cable Books, Volume I. The Community Medium.* Arlington, Virginia: The Cable Television Information Center, 1982.

Jesuale, Nancy, ed., with Smith, Ralph Lee. *Cable Television Information Center Cable Books, Volume II. A Guide for Local Policy.* Arlington, Virginia: The Cable Television Information Center, 1982.

Larrett, R., ed. *Inside Videotex.* Toronto: Infomart, 1981.

Le Duc, Don. *Cable Television and the FCC.* Philadelphia: Temple University Press, 1973.

Lucus, W.A.; Heald, K.A.; and Bazemore, J.S. *The Spartanburg Interactive Cable Experiments in Home Education.* Pub. No. R-2271-NSF. Santa Monica: The Rand Corporation, 1979.

Martin, James, and Butler, David. *Viewdata and Information Society.* Englewood Cliffs, N.J.: Prentice-Hall, 1981.

Michigan State Senate. *Citizens Guide to Cable Television Franchising.* Write: Senator Kerry Kammer, Capitol Building, Lansing, Michigan 46909. (517) 373-2417.

Minnesota Council on Quality of Education Study. *Surveying Need for Low Power Television (LPTV) in Small Rural School Districts.* Minnesota: Educational Management Services, 1982.

Moss, Mitchell L., ed. *Two-way Cable Television: An Evaluation of Community Uses in Reading, Pennsylvania.* Volume I. Sponsored by the National Science Foundation. New York: New York University, 1978.

Muth, Thomas A. *State Interest in Cable Communications.* New York: Arno Press, 1979.

National Federation of Local Cable Programmers. *The Cable Franchising Primer.* NFLCP, 1980.

New Technology Resource Guide. New York: Education Division, WNET/THIRTEEN, 1980.

New York State Cable Commission. *Cable Television Franchising Workbook.* Write: Tower Building, 21st Floor, Empire State Plaza, Albany, New York 12223.

Orton, Barry, ed. *Cable Television and the Cities: Local Regulation and Municipal Uses.* University of Wisconsin-Extension, 1980.

Parker, Lorne A., and Olgrem, Christine H. *Teleconferencing and Interactive Media.* Madison, Wisconsin: University of Wisconsin Center for Interactive Programs, 1980.

Pennsylvania Local Government Commission. *Cable Television in the Commonwealth of Pennsylvania: Analysis and Recommendations.* Harrisburg, Pennsylvania: Commonwealth of Pennsylvania, Local Government Commission, 1979.

Ronka, Robert. "Cable TV: Preserving Public Access." Los Angeles Law Review 4 (1981): 8-13.

Smith, Ralph Lee and Gallagher, Raymond B. *The Emergence of Pay Cable Television: Vols. I-IV.* Cambridge: Technology and Economics, Inc., 1980.

Stearns, Jennifer. *A Short Course in Cable.* 6th ed. St. Louis: Office of Communication, United Church of Christ, 1981.

Technical Papers, NCTA 31st Annual Convention and Exposition—Cable '82. Write: NCTA, Science and Technology Department, 1724 Massachusetts Ave., N.W., Washington, D.C. 20036.

Trempeleau County Kellogg Project: "A Rural Telecommunications Service System." Trempeleau County, Wisconsin: Western Wisconsin Communications Cooperative.

Yin, Robert K. *Cable Television: Applications for Municipal Services.* Report No. R-1140-NSF. Santa Monica: The Rand Corporation, 1973.

Periodicals and Reference Books

Access. National Citizens' Committee for Broadcasting, P.O. Box 12038, Washington, D.C. 20005.

Broadcasting/Cablecasting Yearbook. Broadcasting Publications, Inc., 1735 DeSales Street, N.W., Washington, D.C. 20036.

CableAge. Television Editorial Corporation, 1270 Avenue of the Americas, New York, New York 10020.

Cablefile. Titsch Publishing Company, 1139 Delaware Plaza, P.O. Box 4305, Denver, Colorado 80205.

The Cable TV Databook. Paul Kagan Associates, 26386 Carmel Rancho Lane, Carmel, California 93923.

Cablevision. Titsch Publishing Company, 1130 Delaware Plaza, P.O. Box 4305, Denver, Colorado 80204.

Channels of Communication. 1515 Broadway, New York, New York, 10036.

Community Television Review. National Federation of Local Cable Programmers, 906 Pennsylvania Avenue, S.E., Washington, D.C. 20002.

CTIC CableReports. Cable Television Information Center, 1800 North Kent Street, Arlington, Virginia 22209.

LoPower Community TV. 7432 East Diamond Street, Scottsdale, Arizona 85257.

LPTV Currents. 11800 Sunrise Valley Drive, Reston, Virginia 22091.

LPTV Reporter. P.O. Box 33128, Washington, D.C. 20033.

Multichannel News. Fairchild Publications, 1762-66 Emerson Street, Denver, Colorado 80218.

Television Factbook. Television Digest, Inc., 1836 Jefferson Place, N.W., Washington, D.C. 20036.

Organizations

The Association of Public Power. 2301 M Street, N.W., Washington, D.C. 20037. (202) 775-8300.

Board of Cooperative Educational Services (BOCES). New York State Educational Department, Bureau of Educational Communications, Room 325, Educational Building, Albany, New York 12234. Contact: Bill Humphrey.

The Cable Television Information Center. 1800 North Kent Street, Suite 1007, Arlington, Virginia 22209. (703) 528-6836.

Corporation for Public Broadcasting. 1111 16th Street, N.W., Washington, D.C. 20036.

Federal Communications Commission, Complaints and Information Branch. Cable Television Bureau, Federal Communications Commission, Washington, D.C. 20554.

The International City Management Association Cable Committee. 1120 G Street, N.W., Washington, D.C. 20005. (202) 626-4600.

The National Association of Telecommunications Officers and Advisors. 1301 Pennsylvania Avenue, N.W., Washington, D.C. 20004. (202) 626-3020.

The National Consumer Cooperative Bank. 2001 S Street, N.W., Washington, D.C. 20009. (202) 673-4300 or (202) 673-4334.

The National Federation of Local Cable Programmers. 906 Pennsylvania Avenue, S.E., Washington, D.C. 20002. (202) 544-7272.

The National League of Cities Cable Committee. 1301 Pennsylvania Avenue, N.W., Washington, D.C. 20004. (202) 626-3020.

National Translator Association. Suite 2100, 36 South State Street, Salt Lake City, Utah, 84147.

State of Alaska, Division of Telecommunications Systems. 5900 East Tudor Road, Anchorage, Alaska 99507. Contact: Mel Hoversten, Director.

cable/broadband communications books*
from Communications Press, Inc.

The Cable/Broadband Communications Book series, edited by Mary Louise Hollowell.

Volume 1, 1977-1978 (ISBN 0-89461-027-9), fall 1977.

Volume 2, 1980-1981 (ISBN 0-89461-031-7), fall 1980.

Volume 3, 1982-1983 (ISBN 0-89461-035-X), Jan. 1983.

—and the series' predecessor, *Cable Handbook 1975-1976*, also edited by Mary Louise Hollowell (ISBN 0-89461-000-7), spring 1975.

Creating Original Programming for Cable TV, edited by Wm. Drew Shaffer and Richard Wheelwright, for National Federation of Local Cable Programmers (ISBN 0-89461-036-8), Jan. 1983.

Cable TV Renewals and Refranchising, edited by Jean Rice (ISBN 0-89461-037-6), Jan. 1983.

The New Communicators—A Guide to Community Programming, by Dorothy Forbes and Sanderson Layng (ISBN 0-89461-030-9), 1978.

Communications Technologies in Higher Education—22 Profiles, edited by Ruth Weinstock (ISBN 0-89461-025-2, hardcover; ISBN 0-89461-026-0, paperback), summer 1977.

*Paperback editions, except where otherwise indicated. This list is current as of November 1982. Inquiries about these and/or forthcoming titles should be directed to Communications Press, Inc., 1346 Connecticut Avenue, N.W., Washington, D.C. 20036.